Zhongchao Tan

Academic Writing for Engineering Publication

Guidelines for Non-native English Speakers

Zhongchao Tan
Department of Mechanical and Mechatronics Engineering
University of Waterloo
Canada

CANAPRIL SOLUTIONS
Professional Technical Services

ISBN: 9798644399390 (soft cover)

To My Mother

With love

ACKNOWLEDGEMENTS

I would like to thank my beloved wife and daughter, who designed the lovely book cover and layout. I also want to thank the publishers for their permission to use the materials cited in this book, as well as those giving me permission to reprint materials I had published elsewhere, and most importantly, you, the reader, for reading this book!

CONTENTS

1 Preface

1.1 About the Book

Writing is a type of communication, an important one. Written communication is more than fluent speaking or a good command of grammar, spelling, and punctuation. Writing is a complex task that requires training and practice of many techniques, such as organizing ideas logically, constructing sentences and paragraphs coherently, presenting with appropriate tones, formatting in a stylish manner, and executing in an ethical and professional matter.

Written communication reveals our intelligence of thinking, ability of using words, level of education, and so forth. Good writers are usually creative people with brilliant ideas, which can help capable writers excel in their career development. Regardless of your jobs, you might need to write daily. Writing is a great way to extend your voices that conveys your thoughts and ideas to many people in the world.

Academic writing is also referred to as scholarly writing. It is the writing produced as part of academic works; it can be an article, a book, a report, a thesis, or the like. Academic works, which are primarily produced by researchers and graduate students, are shared with other professionals in their areas of research. Engineering academic writing is one type of technical writing produced by authors in engineering.

This book is aimed at non-native English writers such as international students, as well as researchers who are studying and working in English speaking countries. This book has more than 200 examples for contrast and comparison, a long list of 65 pairs of confusing words, and an emphasis on some essential cultural difference for non-native English writers. Furthermore, an in-depth introduction of copyright and plagiarism at the beginning can help a writer avoid unnecessary challenges from their readers or the copyright holders.

1.2 Scope and Readers

Writing in English can be formal or informal. Formal writing in English is used by students in academics, professionals in Engineering industries, *etc.* It is expected to be professional and should follow rules of grammar, spelling, logic, *etc.* Informal writing does not have to rigorously follow grammatical criteria and it is often for local usages by a particular ethnic or social group in a casual setting. Most informal writing tends to be vivid, colorful, and impressive, but its usefulness is limited to certain contexts because of its use of dialects, slangs, or jargons.

This book focuses on formal academic writing in a professional language and frame. It is written in standard English and provide useful guidelines on development of thoughts, organization with logics, structure in formal style, and usage of the right words. It also pays attention to details such as punctuation, numbers, and date. Informal writing is excluded from the scope of this practical guideline.

This book is designed for non-native English speakers who have to write scientific research articles, technical reports, proposals, engineering thesis, academic books, and other technical documents in English. Many books are available for the training of technical writing; some are free online, and others are priced for sales. However, there are not many books dedicated to non-native speakers of English. Most international students in English speaking countries are capable of reading and understanding other authors' writing. As an engineering professor, however, I have seen and helped many students who need systematic training in writing for their course projects, conference papers, journal articles, and theses. The guidelines in this book have been proven useful to advancing their formal writing skills.

Readers of this book are not limited to non-native English speakers. Students can use the book as self-study materials to improve their communication skills. Engineering instructors can use the book for writing skills development. Engineering students may find this book well balanced between complicated theory and simplicity. Researchers and professionals in engineering industries may find valuable writing techniques in the book too. It is useful to new or experienced writers. With the guidelines in this book, they both can further convey complex technical information with clarity and conciseness.

Someone who dreams of becoming a novelist is not recommended to use this book.

1.3 Book Organization

In this book, the approach to writing is presented in the order of time following typical writing sequence. Successful writing starts with preparation and an outline, continues with drafting and revision, and ends with formatting and proofreading. Admittedly, they often relate to each other and overlap without a clear boundary. One step upstream may affect subsequent steps. For example, when you are working on the outline, you may decide to expand scope of the work. It is difficult to quantify the time required for each step, but in general, more time is needed for a writing task with a greater complexity. By the way, dividing the writing process into different steps may be important to collaborative writing for multiple authors. However, collaborative writing is out of the scope of this book.

This book is organized into three divisions: preparation and outline; drafting and revision; formatting and proofreading. Each division consists of multiple chapters as seen in the table of contents above. The following is a brief chart of these three parts:

Part I. Prewriting
- 2. Ethics and Professionalism
- 3. Outlining

Part II. Drafting and Revision
- 4. First Draft
- 5. Paragraphs
- 6. Sentences
- 7. Words and Phrases
- 8. Punctuation

Part III. Finalizing
- 9. Final Formatting
- 10. Proofreading and Others (*e.g.* copyright permissions)

As indicated by its heading title, Part I prepares you for writing. It introduces ethics and professionalism followed by outlining, which is a tool for ideas organization. Outlining is the first step in writing techniques; a well-developed outline sets up the main stage for writing.

Part II focuses on the process of drafting and techniques of editing. You fill in the outline with organized paragraphs, which is a logical presentation of sentences. Clauses, phrases and words are the building blocks of sentences throughout writing. The first draft precedes editing, when you pay attention to structures of paragraph and sentence, orders of presentations, format, *etc.* This section also includes some language-related aspects like the rules of punctuation, wording and grammar. Revisions are time consuming, but you have to do it again and again until you are satisfied before you share the work with your fellow professionals.

Part III introduces the last steps in finalizing your manuscripts. They include editing and proofreading of the document. Last, but not the least, it also covers the best practices that may help you avoid plagiarism and copyright challenges.

To be practical and useful to the intended readers, theoretical writing skills are presented for the sake of simplicity by using examples - correct ones selected from a variety of authentic texts and incorrect ones from different sources. The examples, which are numbered in sequence and are shaded, stand out from the text. This layout allows intermediate writers to improve writing skills quickly by focusing on the examples.

2 Ethics and Professionalism

Writing ethics demands avoiding plagiarism and resisting the impulse to make false claims. It is critically important to scholars and the readers because their published works may influence many others in the world. Ethical writing also refuses misleading and confusing information or conclusions. These actions can be grouped as professionalisms. They are introduced in this chapter.

2.1 Plagiarism

Plagiarism is not acceptable in the academic world and many sectors including, but are not limited to, higher education, engineering research, and professional publication. Plagiarism may lead serious professional and legal consequences to the individuals who plagiarize. Their employers are often impacted too.

Plagiarism can be avoided by giving appropriate credit to others when you use their unique ideas or quote their exact words. In technical writing, you can give credit to the original creators of the unique ideas with appropriate in-text citations and references. Paraphrased materials should also be credited if the unique ideas belong to others. You also need written permission (with or without payment) from the copyright holders.

Paraphrasing common knowledge does not need a reference. Common knowledge is widely known to the public or understood by your target readers. For example, Newton's second law is common knowledge and is found in all creditable engineering dynamics textbooks. If you are not sure, citations protect you.

2.2 Copyrights

Materials, especially visuals, under the protection of copyrights cannot be published without written permission from the copyright holders. You need to contact the owners for permission and keep the written permission on file for future record. Copyright holders are not necessarily the creators. For example, most authors sign legal agreements to transfer their rights to the publishers of their journal articles or books. In this case, the publishers own the copyrights. You may request and receive permissions to use the materials from the copyright holders, but you still need to give credit to the original creators of the ideas.

2.3 Professionalism

The goal of technical writing is to share knowledge for the greater good of society. The ideas shall be clear to readers. Expository writing is required for engineering publication. It aims to achieve clarity and accuracy using concise language. The writers do not leave it to the readers to analyze and draw conclusions based on their imagination or limited knowledge. You must have a thorough understanding of the subject in order to use expository writing effectively. The following sections are important to professional communication.

2.3.1 Avoiding logical errors

Logic is essential to convincing argument and valid conclusions. Drawing an illogical conclusion is considered unethical because it misleads the readers. It can undermine the writers' credibility and may even have a negative impact on their professional career development. The following are typical logic errors in academic writing.

- A statement lacks reasoning when it is contrary to common sense. (If you write down "a helium balloon falls down", then the sentence should be followed by a good explanation.)
- An over-inclusive statement. It generalizes an observation which is applicable to only a small group. (Some students tend to claim that their computational models are the *best*.)
- A mis-linked statement. It is connected to a previous statement with a logic gap.
- A questionable cause. A logical error as a result of hasty conclusions without examining the actual causes (for example, I felt stomach-ache after eating at a restaurant, so the food in that restaurant must have made me sick.)

2.3.2 Avoiding biased evidence

Never make false claims. To draw a conclusion from omitted or incomplete data, questionable sources, or selected evidence is not only illogical but also unethical. It is unethical, even illegal, to report false, fabricated, or plagiarized results in engineering publication. Do not draw a conclusion from partial information, if you know there is more. For example, drawing a conclusion with selected experimental results is misleading because that is valid only under a narrow range of conditions. It is your obligation to correct any misrepresentations of fact before, and after, publication. Make sure you can stand behind what you write.

The following techniques can help you avoid biased actions.

- Present facts, including verifiable data or statements from creditable sources, instead of subjective opinions. (*See* 4.17.1)

  Make sure that the data support your conclusions. Allow your readers to validate your conclusions from the facts, statistical analyses, examples, *etc.* that you present.
- Present ideas with controlled pace to your readers.

  Assume your readers are new to the subject, although you might be the expert.
- Present negative facts or conclusions as they are.

  Understanding the limitations of your work indicates your qualification and expertise in the field of research.

Visuals can also be misleading when information is selectively omitted or distorted. Distortion can be created by using wrong scales or selected data. *See* 5.4 and 9.10 for more information.

Avoid deceptive language in your writing. It is unethical to deceive or mislead your readers using jargons, euphemisms, or ambiguous words with multiple meanings. Local language should not be used in engineering publication.

2.3.3 Avoiding ambiguity

It is important to avoid ambiguity and misleading information. Words without a clear focus should not be used in technical writing. Fellow professionals expect your communication to be clear and accurate, although it is not always the fact in reality. For example, words like *important, good, well, bad* and *thing* are subjective with multiple meanings and interpretations. Instead, use concrete words and choose the right ones (*see also* Words and Phrases).

2.3.4 Avoiding jargons

Jargons are specialized words or expressions that are used within a profession or group; it is difficult for others to understand. Jargon is understood only by the readers who are unique to an occupational. Jargon may be efficient for communication among insiders, but it should be avoided in technical publications that are aimed at international readers.

2.3.5 Privacy and confidentiality

Private information must be protected according to privacy standards in some countries. Ensure that your writing for publication follows both local and international regulations. Consult with your superior if you are in doubt.

Avoid plagiarism by giving credit to the photographers and securing permission to use them. Photographs are often used to show the appearance and the size of an object. Without permission from the copyright holders, however, they cannot disclose the confidential workings of a mechanism. Such details are normally represented in a schematic diagram or drawing. (*See also* 5.4.2.7)

2.4 Authors and Contributors

Avoid free riders in authorship. Academic integrity affects you, your colleagues, your readers, and many others. Average authorship has grown from three or four co-authors to six or more over the last 20 years. It has caused confusion over accountability and entitlements among academics and professionals. Regardless of the motivations, an unnecessarily long list of co-authors calls for negative perception. (*See* 4.7).

It is important to understand the difference between authors and contributors. Wrong authorship often treats contributors as co-authors. The contributors may have enabled the research by providing valuable resources or administrative support, but they do not participate in the research work or writing. Take authorship seriously.

2.5 Global Communication

It is essential for a non-native speaker of English to recognize that the readers are from a variety of cultural backgrounds, and that they could be anywhere in the world. Acknowledgement of diversity and recognition of cultural difference are key to effective communication. It offers the opportunities for you to reach more readers and to create a strong impact on the society.

In addition to cultural diversity, modern society calls for inclusivity. Gender inclusivity affects the grammatical agreement of your writing (See 6.1.1.3). Words also matter to the ethnic groups, nationalities, religions, and so forth. Use politically neutral words and tones to avoid misunderstanding.

Visuals like graphs, images, and even colors require careful attention. Make sure that visuals are presented with professionalism. Always create simple visuals with consistent labels in the same writing. Avoid unnecessary complication. (*See* Section 5.4, Visuals for more information.)

There are many other writing techniques that have cultural implication, such as positive voice and direct statements. They will be emphasized as they appear in the text throughout the book. All in all, keep your readers in mind when you write; be culture sensitive.

3 Outline

3.1 Preparation

Like any professional tasks, success in writing requires solid preparation. Preparation for writing is as important as drafting and revising the document. You can accomplish a thorough preparation by understanding the following factors.

1. The purpose of your writing
2. The primary readers of your document
3. The scope of your work
4. The target platform of publication

3.1.1 Purpose of writing

There are many types of academic writing, such as report, article, thesis, and similar documents. Academic journals are often referred to as *Periodicals* in a library. There are many varieties of books, each contains multiple chapters with a coherent focus. Conference papers refer to articles that are written for the conferences, where you can present your results to the community. A conference paper is normally shorter than a research article. A typical research article has thousands of words. A dissertation is a degree-requirement work written by a doctoral student. It may be as short as 50 pages or as long as 1000 pages. It takes years for a doctoral student to complete the dissertation; conference papers and peer reviewed articles may be based on the contents of the dissertation. Although they have different sections, most of the sections have common organization and structures.

All writing has a purpose. Ask yourself why you write before you begin. Engineering academic programs normally require writing of course projects, theses, and dissertations, but they do not dictate in publication of articles or books. There are many reasons to write for publication, and only write when you know the exact answers.

Although the guidelines in this book are aimed at academic writing, basic writing principles often apply to similar documents such as progress reports, lab reports, investigative reports, test reports, white papers, and course project reports.

Two most important academic writing in engineering fields are research articles and thesis or dissertation. Research articles published in engineering periodicals aim to further the knowledge and to report the advances in a specialized field. The readers are normally professionals, such as academics, educators, engineers, graduate students, and scientists in the world, who also regularly contribute articles to those journals. Articles published in journals, especially those prominent international journals, may raise the profile of your employer and improve your chances for professional advancement.

There are several golden standards for publications in prominent international journals. First, your work (knowledge or approaches) must be original. The significance of your contributions justifies the time and effort to write an article, and the time of others for them to review and publish the knowledge globally. It is a normal requirement that the work in writing matches the scope of the journal to be submitted for consideration of publication. Other factors such as the prominence of the journal, review time, and frequency of citation should be considered too; by this time, you are close to finishing your writing.

3.1.2 Identifying the readers

It is important to identify your readers before you begin writing. Who are your readers? Ask yourself the following questions before writing down anything. Precise answers to these questions will determine your scope of writing, and the balance between clarity and conciseness.

- What do my readers already know about the subject?
- How much background information do I need to put down in writing?
- Should I define basic terminology?
- Am I communicating with international readers?
- Are my readers qualified professionals in the scope of the work?

Keeping your readers in mind is crucial to effective communication with them. Consider your readers' levels of knowledge. Many publications are for readers with diverse background. You need to accommodate their needs as much as you can. Most engineering publications are aimed at international readers from all over the world, within your professional community. Admittedly, it is challenging to satisfy all the readers with different needs. Then you may have to determine your primary readers, such as the students in your professional community, your peer researchers in your research areas, or those who may make decisions based on your writing.

In addition, you ought to understand their cultural values that underlie the language to be used for international readers. In a globalized world like today. You are urged to write with clear and complete sentences. You should avoid the following writing mistakes, which may confuse international readers.

- Overly simplified style
- Unusual word order or rambling sentences
- Informal words for humor, irony, and sarcasm
- Localism, jargon, and slangs
- Abbreviation, terminology, *etc.* without definition first

Regardless of your readers, any clear writing in English is expected to be concise and coherent with clarity. Typical technical writing is for educated readers, including university students, researchers, and professionals in specialized industries. This type of highly educated readers determine how you write; you need to make the following decisions accordingly.

- Provide enough background information with details.
- Provide headings and subheadings using words in the field of primary readers.
- Use plain and technical vocabulary in body text.
- Write a procedure with details so that readers can follow easily.
- Use clear visuals besides written text for clarity.

3.1.3 Scope of writing

The scope should be determined before drafting the document, although you may refine it at the later stages such as drafting and editing. A well-defined scope saves you time in writing.

Effective writing has a clear focus. Avoid overly broad objectives (*see* Example 3-1). Simply ask yourself what your contributions, and only yours, are then you can establish your primary scope of writing. Be precise when you state the objectives, which are normally justified by state-of-the-art background review. *See* 4.17.

Example 3-1. Avoid overly broad objectives

Incorrect: The objective of this research is to develop a novel CFD model.

Correct: The objective of this thesis research is to develop a drift-flux model for indoor aerosol dispersion in classrooms.

Comments:

Many researchers have developed or are developing novel CFD models for a variety of applications. It is unlikely for one to fully address the complexity. You need to be specific on the questions to be answered or the problem to solved by *your* work being presented.

3.2 Organization of Writing

The organization of engineering academic writing depends on the length of the document. Two most important types of academic writing are research articles (short) and theses or books (long). They follow similar principle and writing styles. Bearing in mind that the structure is flexible and can vary with journal. Their common features are introduced as follow.

Table 3-1 shows typical divisions that technical scholarly writing may have in the order they typically appear. You have more freedom in organizing books and theses than conference papers and journal articles. Most periodicals and conference organizer use their own prescribed organizations for the major sections defined by level-one headings. Almost all journals furnish guidelines for writers to follow and make them accessible on the journal websites. The common level-one heading in most international journals include *Title*, *Authorship*, *Abstract*, *Introduction*, *Methodology*, *Results and Discussion*, *Conclusions*, and *References*. Optional but valuable sections include *Recommendations*, *Limitations*, and *Acknowledgments*. Some journals, because of page limit, may ask you to include Supplemental Materials, which are not printed out in hardcopy, but will be accessible online.

Table 3-1. Typical sections in engineering publication

Group	Section (Heading 1)	Long document	Short document
Front Matter	Title	Yes	Yes
	Author(s) and Affiliation(s)	Yes	Yes
	Abstract	No	Yes
	Executive Summary	Yes	No
	Table of Contents	Yes	No
	List of Figures	Yes	No
	List of Tables	Yes	No
	Foreword	Optional	No
	Preface	Optional	No
	List of Abbreviations, Acronyms, and Symbols	Yes	Optional
Body	Introduction	Yes	Yes
	Chapters	Yes	No
	Methodology	Yes	Yes
	Results	Yes	Yes
	Discussion	Yes	Yes
	Conclusions	Yes	Yes
	Recommendations	Yes	Optional
	Explanatory Notes	Optional	No
	Acknowledgments	As needed	As needed
	References	Yes	Yes
Back Matter	Appendices	Yes	As needed
	Bibliography	As needed	No
	Glossary	As needed	As needed
	Index	Recommended	No
	Supplemental Materials	No	As needed

However, it does not mean all of them will appear in one single document. Long documents normally have more sections (or divisions) than the short documents do. You need to check with the publisher for the prescribed organization and sections and ensure that you have the freedom to create your own sections. Despite the variation of sections and elements in engineering academic writing, they share the same writing principles. They are introduced as follows.

3.2.1 Front matter

The front matter describes the general ideas of the work and summarizes the sections of the document. Although not all academic writing includes every element of front matter listed in Table 3-1, the title, author(s) and affiliation(s) are normally essential to all documents. The rests are optional, and they are used as needed. For example, *Abstract* is usually mandatory for journal articles and dissertations, and *Forward* or *Preface* for books.

3.2.2 Body

The document body contains the background information (introduction) for the work, detailed state of the art, procedures followed to conduct the research, data analysis information, results obtained, and conclusions drawn from the results. *Recommendations* and limitations of the work are made clear, if appropriate. All this information can be enhanced using visuals like figures and tables.

The Introduction should be interesting to your readers. It explains the state of the art of the work, the knowledge gaps to be addressed, and the problems that need to be solved through *your* work. The *Introduction* justifies the work by answering the questions of why your contributions are important and new. It typically ends with well defined objectives. (*See* 4.17)

The body of your document follows the statement of the objectives. It begins with a methodology, which describes the procedures of obtaining your own data, if any, in addition to data in literature. If the approach to research is primarily experimental, the *Methodology* section should be written with enough details allowing readers to reproduce the data following the prescribed procedure. At the minimum, readers have the right to know the experimental procedure, raw data collection method, and expected results to support the conclusions. They all support your argument.

The *Results and Discussion* are presented after the Methodology, and the *Conclusions* summarize the key contributions of your work. Avoid the biased opinions toward your own conclusions.

In attempt to establish a logical argument, you also need to draw information and data from external sources. They must be credited by references and in-text citations to avoid plagiarism. The *references* are listed after the body text. Visuals are also used to reinforce your argument, and they should be presented following the appropriate style and ethics. (*See* Visuals)

3.2.3 Back matter

The *back matter* contains additional and supplementary materials, which are useful to the readers for various reasons. *Bibliography*, which is different from the references, provides the readers with additional information that concerns most readers. *Appendices* (to long documents) or *Supplemental Materials* (in articles) expand on the subjects that might be distractive to the primary readers if they were included in the main body, but they are useful to others who are interested in a better understanding of the subjects. *Glossary* and *indexes* are for multi-chapter and multi-volume documents such as books, dissertations, and encyclopedia. They are useful tools for the readers to locate relevant information in the main body.

3.2.4 Headers and footers

Headers appear at the top of the designated pages; footers appear at the bottom of the pages. The text in headers and footers depends on the publication, and there is enough room for the authors to show their creativity. The headers may include the document title, the section title, and the footers, other information to help readers track their reading progress. Either the headers or the footers should include the page numbers, more often as footers than headers. Be careful not to crowd your headers and footers to avoid visual clutter. Finally, information included in headers or footers should be concise and usually not meant to distract readers from the main text.

3.3 Outline

A well-developed outline and a logical order of presentation help you bring coherence and unity to writing. You will follow the outline when you begin the first draft. Like a road map, the outline keeps your ideas focused on the purpose of writing so that you can logically draw your conclusion. A logical organization ensures that your materials are presented in a coherent manner. It also enables the clarity, unity, and smooth transition from one text to the next without omitting important details.

There are two steps for the creation of outlines: a short topic outline followed by a lengthy outline with sentences. The former consists of short phrases that reflect the order of presentation. The latter extends short phrases into complete sentences. Each sentence in the outline states the controlling idea of the corresponding paragraph in the draft. The topic sentences of paragraphs in the draft ensure that your writing is well organized. (*See* 5.1. Paragraph Structure)

The outline of an engineering document ensures that your writing begins with introduction, followed by the main body, and ends with the convincing conclusions. It begins with headings, lists, and other special design features for the framework for your writing. At this point, you must begin to consider the scope of your writing.

This is also a good time to consider where visuals (such as figures and tables) to be positioned for effective presentations. You should use visuals to achieve clarity and smoothness in writing. In your outline, simply reserve the space by writing "Visual for . . .," noting the purpose of the figure, table, or another type of visuals. When you begin drafting, integrate the visuals smoothly with your text (*See* Section 5.11 Visuals).

3.3.1 Creating a short outline

Outlining is the critical step to the development and organization of your ideas. An outline will not confine your thoughts; it's there to guide you through drafting. The function of an outline in writing is like the framing of a house; a constructor cannot build a functional house without the clear blueprint (framing work) of the floor plan of rooms and the fixture in the rooms. You can still change the contents when you are writing. An outline is only a non-restrictive guideline. Outline is meant to make your writing easy, but it shall not dictate your flow of ideas or choices of words.

You normally do not have the opportunities to observe how outlines are created and revised by reading the final version of the documents. Many engineering programs do not offer academic writing courses to graduate students; most of them are trained by their academic supervisors. As a result, academic writing skills become personal and is often omitted out engineering education. You can use the following step-by-step guidelines for outlining.

Step 1: List major headings

For example, the major headings of a typical international journal article consist four levels at the minimum as follows. (There are normally more than four level-1 headings.)

- Introduction
- Methodology
- Results and discussion
- Conclusion

Step 2: Establish the minor headings within each major heading

Arrange your minor points under their major headings. Often you will need more than two levels of headings, especially for complex subjects. It is normal to use Heading 3 and Heading 4 to better organize the writing in proper relationship. It is unusual for the headings to go beyond Heading 4; otherwise, the numbers become cumbersome.

Visuals are an integral part of your outline and should be put where they are likely to appear. Describe each visual with a tentative caption, *e.g.*, "schematic diagram of…" All these are temporary and evolving like other information in the outline. You can add or remove visuals as needed.

At this step, make sure that each of your headings is marked with the appropriate sequential Arabic numbers combined with optional letters. You can cross-reference the decimal numbering system while you are developing the document. In addition, make sure that the headings have parallel structures.

Example 3-2 shows a sample short outline with three levels of headings. It has only Arabic numbers, but you can use letters as you prefer.

Example 3-2. Short outline with headings, phrases and visual titles

1. Introduction
    1.1 Literature review
        1.1.1 Motivation
            Air pollution, combustion, air cleaning…
        1.1.2 Experimental works
            Adsorption, absorption, thermal conversion…
        1.1.3 Numerical works
    1.2 Knowledge gaps
        Cost-effectiveness, technical feasibility…
    1.3 Objectives
        Kinetics, mechanism
    1.4 Outline of the thesis
2. Methodology
    2.1 Overall experimental setup
            Figure. Schematic diagram of the setup
    2.2 Test apparatus
            Figure. Photo of the test apparatus
    2.3 Instrumentation and data collection
            Table: List of devices and data collection
    2.4 Data analyses and expected results
3. Results and Discussion
4. Conclusions

Step 3: Create the lengthy outline with sentences

You may consider this step as the beginning of rough drafting. As an important step in writing, it is elaborated in the next section under a level-three heading.

3.3.2 Creating a lengthy outline with sentences

What you have now is a short outline, and it can be extended into a lengthy outline with complete sentences and paragraphs without precise grammar, language, or punctuation; they will be taken care of when you revise the draft following the guidelines about paragraphs, sentences, phrases, words, and even punctuation. This conversion can be part of the drafting. However, a sentence outline is recommended because it may be the most difficult part of your writing. A complete lengthy sentence outline is almost like a *crude* draft (Example 3-3). The topic or sentence outline should be flexible because it may change when you write the first draft.

Example 3-3. Lengthy outline with topic sentences

1. Introduction
    - 1.1 Literature Review
        - 1.1.1 Motivation
            - Air pollution impacts human health and the environment.
            - Fossil fuel combustion is one of the sources of air emissions.
        - 1.1.2 Experimental works
            - There are several technologies available for air emission control. Selective catalytic reduction; wet scrubbing; low NOx combustion.
            - (Visuals to show experimental setup)
            - There is a need of cost-effective technologies for nitric oxide (NOx) emission control.
            - This work is focused on NOx emission control by wet scrubbing using cobalt-based solvents.
            - (Table to compare the pros and cons of earlier works)
        - 1.1.3 Numerical works

[Repeat the same practice under other headings]

Formatting the headings is optional, even though you might think it helps track their relative importance when you write the first draft. Actually, formatting any part of the document at this stage is a waste of your time because you may change your mind while you are drafting. Headings are added, removed, or moved while you are working on the first draft. Therefore, you should format your document after finishing all revisions. (*See* 9 Final Formatting)

~~~

With the established purpose, readers, scope, and outline, you are ready to write the first draft by expanding the outline into sentences and paragraphs. Meanwhile, you create visuals and integrate them into the text. The next chapter introduces the practical techniques for drafting.
~~~

4 First Draft

4.1 Draft Procedure

Your first draft follows the lengthy outline with sentences. The headings in the outline divide your document into sections. These sections may be long or short, depending on their functions in your document. Each of the sentences should become a paragraph following the paragraph structures (*See* 5.1). All sentences in the paragraphs should follow sentence structures (*See* 6.1).

An experienced writer begins drafting with the *Body* of the document. Consider writing the sections in the *Front Matter* last when you will have a clean draft. Introduction serves as a frame into which readers understand the background, the motivation, and the overview of the work. *Conclusions* tie the main objectives together; your readers should be convinced that you have achieved the *Objectives* as stated at the end of *Introduction*. Also consider including *Recommendations* to your readers, such as a course of action for further research, a judgment call, or merely a reiteration of your main points.

When you write the first draft, you can pay little attention to the details like punctuation, grammar, and spelling. You might imagine that you were talking to the readers who are sitting in front of you. They would tolerate and understand even if you do not choose the exact words in verbal communication. Focus on the important matters by following the ideas established in the outline as it serves as a road map for the first, rough draft.

Language is important to your final publication, but language errors should not be your concern at the step of drafting. Otherwise, you can not concentrate on the more important task, which is the connection of ideas. You can pay some attention to the details in the second or third revision. Ultimate refinements come with final revision and formatting. Flawlessness can be reached after proofreading.

4.2 Clarity, Coherence and Unity

Clarity is essential to engineering publications. A good understanding of grammar and special terminology enables you to communicate clearly and precisely. However, academic writing for engineering publication takes much more than grammar and spelling. Organizing your ideas in a logical sequence is critical to achieving coherence and unity. You ought to ensure that each paragraph in your draft has a controlling idea supported by a body of evidence.

To achieve clarity, convert each of the sentences in the outline into a paragraph with a single controlling idea. There are various techniques in constructing paragraphs. Typical paragraphs in academic writing consist of three components: *the topic sentence*, *the body sentences*, and *the concluding sentence*, in the order of sequence (*See* Figure 4-1). This paragraph structure ensures one idea in each paragraph.

Statement of the idea
Support evidence 1
Support evidence 2
Support evidence 3
Closing sentence

Figure 4-1. Simple illustration of the paragraph structure

Each sentence plays an important role in the paragraph. The topic sentence is the statement of the controlling idea of the paragraph. The body sentences support the statement using logical arguments, examples, analyses and other information. The paragraph ends with concluding sentences, which remind the readers of the paragraph's main point.

You can further improve the clarity of your writing with plain words, sufficient details, simple definitions, proper emphasis, correct subordination, and smooth transitions. Emphasis and subordination are two complementary techniques that enable you to differentiate the importance of the phrases, clauses, and sentences. Otherwise, a reader may have to guess their importance. Appropriate transition is important to achieving conciseness and coherence in paragraphs. Transition is also essential to the smooth flow from one paragraph to the next or one sentence to another.

Chapters 5-8 introduce the tactics and techniques used by experienced writers. Chapters 5, 6, 7, and 8 focus on various techniques in paragraph construction, sentence variation, word selection, and punctuation usage. Regardless of the elements, engineering writing shares the following common principles.

4.2.1 Writing with plain language

A technical writer usually uses plain language to explain a complex idea, which is a skill that requires systematic training and continual practice. Conciseness is important but clarity is essential to technical writing. Sentences can be shortened by omitting articles, pronouns, or verbs. However, it shall be done without grammatical errors or lose of clarity.

Example 4-1. Confusion caused by an unexpected comma

Incorrect: Acidified rain, ozone, and particulate matter, can all have a devastating impact on the environment.

Correct: Acidified rain, ozone, and particulate matter all can have a major impact on the environment.

4.2.2 Description with details

Details are crucial to the engineering writing aimed at readers that are unfamiliar with the work. Avoid assumption that your readers share the same background information, are capable of data analysis, or can draw right conclusions from the same data. Treat your readers as if they were first year college students.

The key to effective description of a procedure, for example, is the accurate presentation of details. The process description begins with a system and function followed by those components. The descriptions can be complete within one paragraph for a simple system or multiple paragraphs for a complex one, but all descriptions should integrate each component into the function of the system.

Examples for writing with details are available throughout this book. Some may be presented with emphasis. You should be able to master this type of writing skills with continual practice.

4.2.3 Definition of terminologies

Definition is important to clarity and accuracy. Do not assume readers can understand you with what you write, otherwise it causes vagueness and misinterpretation. Defining terms or concepts for your readers is often essential to improve clarity.

Definitions can be accomplished by analogies, by causes, by components, by examples, and by exploration of origins. Giving examples (using *for example*, *for instance*, *such as*, *like*) is the easiest approach. However, terms defined by their causes are especially effective in writing scholarly articles for international publication. Definition by components helps readers to understand a formal definition by breaking the concept into different smaller pieces. Under certain circumstances, exploration of origins may be more effective than others; it is especially useful to terms with unfamiliar *Greek* and *Latin* roots.

Use definition techniques with care when you need them (See Example 4-2). Avoid circular definition (Example 4-3), and avoid "*is when*", "*is where*", and the like in definitions (Example 4-4).

Example 4-2. Definition by components

Formal definition:

Combustion is a chemical reaction between a fuel and an oxidant that produces energy, usually in the form of heat and light, and new chemical species.

Definition by component:

The chemical formula shows that combustion requires fuel and oxygen. Oxygen is normally taken from air; and fuel can be any flammable material. A combustion starts with ignition, and the heat produced sustains the combustion process. Light is generated at the same time. Furthermore, new chemical species are produced by combustion; they are mostly air pollutants or solid wastes.

Example 4-3. Avoid circular definition in writing

Incorrect: Spontaneous human combustion is the combustion of a living human body spontaneously.

Correct: Spontaneous human combustion is the combustion of a living human body without an external source of ignition.

Example 4-4. Avoid "*is when*" style in definition

Incorrect: A supercritical fluid is when the temperature and pressure of any substance is above its critical point.

Correct: A supercritical fluid is any substance at the temperature and pressure above its critical point.

4.2.4 Writing with the right pace

Pace is the speed of presentation. A carefully adjusted pace reduces ambiguity and improves clarity of your writing. Control your pace when you present your ideas to the readers. Pace can be well controlled by general introductory sentences, concise sentences, and emphasis. Hasty pace may occur when you assume that your readers are experts in the field of concern. (Example 4-5).

Example 4-5. Pace of presentation

Hasty: The electrospinning device is powered by a high-voltage power supply and produces nanofibers in the range of 10-800 nm in diameter. It is designed to operate under normal conditions of room temperature and low relative humidity, within a ventilation hood, and may be used for polymer, metal oxides and other materials of similar properties when needed.

Right: The electrospinning device, which is powered by a high-voltage power supply, produces nanofibers in the range of 10-800 nm in diameter. Designed for normal conditions of room temperature and low relative humidity, this device should be used in a ventilation hood. It may be used for polymer, metal oxides, and other materials when needed.

4.2.5 Writing with the right tense

Table 4-1 shows the six tenses and twelve forms of tenses in English. The *simple past tense* is often used in engineering reports, indicating that the research took place in the past. For example, "A thermal couple *was used to* measure the temperature at the center for the furnace."

Table 4-1. Examples of twelve forms of tenses in English

Tense	Basic Form	Progressive Form
Simple future	He will do	He will be doing
Simple present	He does	He is doing
Simple past	He did	He was doing
Future perfect	He will have done	He will have been doing
Present perfect	He has done	He has been doing
Past perfect	He had done	He had been doing

However, the present tense should be used for the following cases (*See* Example 4-6).

- General truths
- Actions without time constraints
- Routine activities that can occur in the past, present, and future
- Authors' opinions
- Contents in dated documents

Example 4-6. Special cases for present tense

The moon orbits the Earth once every 27.322 days. [*General truth*]

It is warmer in summer than winter. [*No time restriction*]

I get up at 8 AM every morning. [*Routine activity*]

In his 1905 paper on "special relativity", Albert Einstein argues that space and time are bound up together, a complicated idea that contradicted the long-held belief in something called ether. [*Author's opinion; contents in the work.*]

The simple future tense expresses that something occurs after the present. It uses phrases like *is going to* or words like *will* before the main verb. The simple future tense can be used for proposal writing, indicating the plan of activities. Proposal writing is beyond the scope of this book, although a good part of this book can be used as guidelines for proposal preparation. Back to engineering publication, use the simple future tense as needed, and try to avoid the unnecessary future tense. (Example 4-7)

Example 4-7. Avoid unnecessary future tense

Incorrect: The model will be explained in Chapter 3.
Correct: The model is explained in Chapter 3.

Incorrect: This literature review will focus on the production of fuel products from carbon dioxide and water.
Correct: This literature review focuses on the production of fuel products from carbon dioxide and water.

A consistent use of tense is important to the clarity of writing (*see* Example 4-8). Be sure to shift tense only when a real change in time is needed. Illogical shifts in tense are confusing to the readers. Imaging that you are describing a procedure for the assembly of a test apparatus, a shift from the past tone to the present one would confuse the readers.

Example 4-8. Consistence in tense

Incorrect: He loads the reactants into the reactor before he turned on the heater.
Correct: He loaded the reactants into the reactor before he turned on the heater.

4.2.6 Point of view

Point of view is another factor affecting the clarity of writing. Point of view can be expressed in first person (*I, we, my, our*), second person (*you, your, yours*), or third person personal (*she, her, it, they*) pronouns. Using consistent point of view in the same sentences helps avoid confusion.

Impersonal point of view is normally used in academic writing for engineering publication. An impersonal point of view is effective in emphasizing the subject matter. The methodology of conducting and presenting scientific and engineering research is well defined by professional societies, fellow researcher, and industrial practitioners. They should not vary with the writer's or the reader's point of view.

Avoid rapid shifting of the point of view to maintain a smooth flow of thoughts. (*see* Example 4-9). An illogical shifting from the third person, which is typical in technical writing, to the first person in mid-sentence will likely confuse many readers. In practice, it is occasionally necessary to use the pronoun in technical writing to avoid awkward sentences, where *one* or *the writers* instead of *I* or *we* is used.

Example 4-9. Point of view and positive tone

Awkward: It is noted that the performance of Sample A is not as good as that of state-of-the-art MEAs reported by Gas. (*ref.* Liu et al. 2009)

Revision: We noted that MEAs (reported by Gas) performed better than Sample A did. *[Such a revision improves conciseness and positiveness.]*

4.2.7 Using positive tones

Like verbal communication, written communication shows your tone too. The overall tone can be positive or negative. Your tone of communication reveals your personality, as well as your general attitude toward the subject matter. A positive tone enhances your image. For this reason, avoid using negative words such as *unfortunately*, *sadly*, *absurd,* and *ridiculous* in professional writing. You can achieve positiveness in writing by transforming the negative into opportunities (Example 4-10). *See also* 6.1.2 for emotional modifiers.

Example 4-10. Transform the negative into opportunity

Negative: Sadly, the conversion efficiencies of these reactors are still too low for them to become a solution for the industry.

Positive: ~~Sadly,~~ The conversion efficiencies of these reactors are still ~~too~~ low and further research is needed for them 9to become a solution for the industry. *[Implies opportunity]*

Readers might view the message as negative when they see the words *not*, *no*, and those with the same function (*e.g.*, *neither/nor*). Minimize the usage of this type of words in your writing and use them only when it is necessary. You can transform a negative tone into a neutral tone by replacing *not* and the word modified by *not* with a prefixed word, phrase or clause of the same meaning (*e.g. impersonal* for *not personal*; *need improvement in accuracy* for *not accurate*). *See* Example 4-11.

Example 4-11. Reduce negative of tones using prefixed words

Negative: Personal point of view is normally not used in academic writing for engineering publication.

Neutral: Impersonal point of view is normally used in academic writing for engineering publication. *[See 4.2.6]*

Grammatical voice and grammatical tone are different, and they can be easily mixed up. With continual practices, however, you will understand their difference and use them properly in writing.

4.2.8 Active and passive voices

A grammatical voice can be *active* or *passive* (Example 4-12). Both active and passive voices are effective, depending on the emphasis. The active voice emphasizes the performer of action; the passive voice emphasizes the subject being acted upon. Make sure that the voices are consistent in the same sentences.

Example 4-12. Active voice and passive voice

Active: When driving a car, the engine pistons turn the crankshaft and drive the powertrain under the car. The tires rotate and move the car forward.

Passive: Gas pressure and temperature should be checked before opening the reactor.

Active Check gas pressure and temperature before opening the reactor.

Comments:

In the passive voice, it is not clear whether the pressure and temperature are checked; the active voice clearly indicates that the performer of the action.

Active voices are highly valued in English. The active voice is generally preferred because it is concise, it improves clarity, and it avoids confusion. In some cases, the passive voice fails to identify the performer. *See* Example 4-13.

Example 4-13. Passive voice leading to vagueness

Passive: Hurrying to finish the data collection on time, two computers were used simultaneously. *[This sentence implies that two computers were hurrying!]*

Active: Hurrying to finish the data collection on time, we used two computers simultaneously *[It is clear in this sentence that the performer of the action is we, the authors.]*

Passive: The computers were turned off before the tests finished.

Active: The lab manager turned off the computers before the tests were finished.

However, it does not mean that the passive voice is useless. The passive voice is effective or necessary when it is unnecessary to identity the performer of the action; it achieves emphasis over the object. For instance, the first sentence in Example 4-14 emphasizes the *abacus* and its value; the inventor or who used it was less important.

Example 4-14. Passive voice used for emphasis on object

The abacus was early used for arithmetic tasks.

This paper was published in 1905.

The passive voice may be appropriate to the explanation of a process or a procedure. This is important to engineering publication because, in many cases, it is not necessary to state who the performer is. The readers should be able to reproduce the work following the procedure. The paragraphs in Example 4-15 are written in passive voices to describe an experimental procedure (Dolan *et al.*, 2010).

Non-native English writers tend to follow the habit of their native languages. The passive voice is used frequently in some languages, but in others, not at all. Japanese, for example, uses the passive voice quite frequently. Remember that using the appropriate grammatical voice is important to the writing in English.

Example 4-15. Passive voice for description of procedures

2. Experimental

Experiments were performed in a batch 69 mL reactor constructed from stainless steel 316 tubing (w17% Cr, w13 Ni, w3% Mb, w2% Mg, <0.03% S), which was heated in a muffle furnace preheated to the set point temperature. The reactor was filled with 5, 10, 15, 25, or 35 mL sodium carbonate solution containing 4% by weight microgranular cellulose (Sigma) and 0.04% by weight 5% Pt/Al_2O_3 (Sigma, powder). The 5, 10, 15, 25 or 35 mL slurries permitted study of headspace fractions of 93, 86, 78, 64 and 49% respectively.

Prior to heating, the reactor headspace was filled and evacuated with argon several times to purge the headspace of molecular oxygen. Finally, the headspace was filled with 15 psi of argon to allow for sufficient pressures for gas sampling in cases where little gas was produced ....

4.2.9 Figure of speech

Figure of speech may improve clarity; they help improve the effectiveness of communication between the writer, who is typically an expert in the scope of writing, and novice readers. To some extent, figures of speech make writing more colorful, interesting, and lively.

Typical figures of speech include *analogies, hyperboles, litotes, metaphors, metonyms, personification,* and *similes.* However, analogies and similes are relevant to technical writing, and the rest are characterized as exaggerations, understatements, emotions, or inaccuracy.

- **Analogies** often use the structure *as* … *as* for comparison.

  Use it carefully in engineering writing, because elaboration is necessary to avoid confusion. (Example 4-16)

- **Similes** are direct comparisons of two objects or concepts using the word *like* or *as.* (Example 4-17)

Example 4-16. Figure of speech: analogy using *as… as*

Outlining in writing is as important as framing in building construction.

Example 4-17. Figure of speech: Simile using *like*

Outlining in writing is like the framing of a house; a constructor cannot build a functional house without a well-structured framing work. *[Elaboration in the second part.]*

Despite the effectiveness, figure of speech should be used only when it is necessary, especially in engineering writing that is aimed at international readers. Non-native English readers may translate figures of speech literally, resulting in misunderstanding of your ideas.

4.2.10 Writing with transition

A well-structured paragraph is characterized with clarity, unity, coherence, and adequate development. *Unity* refers to the singleness of idea as described by the topic sentence: one idea in each paragraph. *Coherence* is the connection of sentences into a logical single point of view by transitional words. *See* Chapters 5 and 6.

4.2.10.1 Transition in sentences

Using the right transitional words (*specifically, in addition, furthermore, however, etc.*) is key to the coherence of the paragraph. An incomplete list of transitional words and phrases can be found in Appendix, Transitional Words and Phrases.

4.2.10.2 Time-order and sequence in paragraphs

Time-order words can be used to improve clarity and coherence by connecting sentences. They can also be combined with lists using numbers or bullet points. Table 4-2 lists some time-order words. These words usually introduce paragraphs, sentences, or items in a list. In addition, easy-to-follow instructions can be described with simple short steps, which can be organized with time-order words. Simple enumeration in the order of sequence (*first*, *second*, *third*, and so on) also provides effective transition between sentences within the paragraph. See Example 4-18 and Example 4-19 (Givehchi, 2015).

Table 4-2. Time-order words used in technical writing

Before	Begin	In between	Last
Earlier	To begin	Next	Finally
In the past	First	Then	In conclusion
Preceding that	To start with	Later	In the end
Previously	At the beginning	Consequently	At the end
Prior to	To begin with	After	At last
Previous to	Initially	Subsequently	Ultimately

Example 4-18. Using time-order words for coherence

Published:

PVA nanofibers were made with different applied voltages, tip to collector distances and deposition times. The morphologies of these electrospun filters were then characterized by SEM images coupled with an automated image analysis method. Using NaCl airborne nanoparticles in the size range of 10-125 nm, the single-layer and multilayer filters were also evaluated in terms of filter quality factor. The effects of the electrospinning parameters on filter quality factor were determined to identify the important factors affecting filtration performance of PVA nanofibrous filters.

Revision:

PVA nanofibers were first made with different applied voltages, tip to collector distances, and deposition times. Then, the morphologies of the electrospun filters were characterized by SEM images coupled with an automated image analysis method. Later, the single-layer and multilayer filters were also evaluated in terms of filter quality factor using NaCl airborne nanoparticles in the size range of 10-125 nm. The effects of the electrospinning parameters on filter quality factor were finally determined to identify the important factors affecting filtration performance of the PVA nanofibrous filters.

Example 4-19. Description of sequence with time-order words

Published:

A custom-made electrospinning setup was used in this study for filter sample preparation. The relative humidity and temperature of the air inside the housing were 39±4% and 23±3°C, respectively. A 5-ml syringe was loaded with a solution of PVA polymer, which has a molecular weight of 89,000-98,000 g·mol-1 (Sigma Aldrich Canada). The desired solution concentration of 10% w/w was prepared by diluting the PVA in distilled water at 90°C and stirring overnight (Appendix B). A 22-gauge stainless steel capillary needle with an inside diameter of 0.413 mm was attached to the syringe. The syringe was mounted on a syringe pump (Kd Scientific), which was used to control the flow rate to 0.3 ml·hr^{-1}. A lab jack was used to adjust the vertical distance between the capillary needle and the grounded collector. A high-voltage power supply (Gamma High Voltage, ES50P-5W) was employed to apply the high voltage between the capillary needle and an aluminum collector.

Revision option 1 - restructuring the sentence:

A custom-made electrospinning device was used in this study for filter sample preparation; the relative humidity and temperature of the air inside the housing were 39±4% and 23±3°C, respectively. Prior to electrospinning, a solution of PVA polymer (with a concentration of 10% w/w) was prepared by diluting the PVA in distilled water at 90°C and stirring overnight (see Appendix B). PVA has a molecular weight of 89,000-98,000 g·mol^{-1} (Sigma Aldrich Canada). To begin with, the solution was loaded into a 5-ml syringe with a 22-gauge stainless-steel capillary needle, which has an inside diameter of 0.413 mm. After that, the syringe was mounted on a syringe pump (Kd Scientific), which was used to control the flow rate to 0.3 ml·hr^{-1}. Subsequently, A lab jack was used to adjust the vertical distance between the capillary needle and the grounded collector. Finally, a high-voltage power supply (Gamma High Voltage, ES50P-5W) was used to apply the high voltage between the capillary needle and an aluminum collector.

Revision option 2 - using a list:

A custom-made electrospinning device was used in this study for filter sample preparation; the relative humidity and temperature of the air inside the housing were 39±4% and 23±3°C, respectively. Samples were prepared following these steps.

- Prior to electrospinning, a solution of PVA polymer (with a concentration of 10% w/w) was prepared by diluting the PVA in distilled water at 90°C and stirring overnight. PVA has a molecular weight of 89,000-98,000 $g\cdot mol^{-1}$ (Sigma Aldrich).
- To begin with, the solution was loaded into a 5-ml syringe with a 22-gauge stainless-steel capillary needle, which has an inside diameter of 0.413 mm.
- After that, the syringe was mounted on a syringe pump (Kd Scientific), which was used to control the flow rate to 0.3 ml/hr.
- Subsequently, A lab jack was used to adjust the vertical distance between the capillary needle and the grounded collector.
- Finally, a high-voltage power supply (Gamma High Voltage, ES50P-5W) was used to apply the high voltage between the capillary needle and an aluminum collector.

Comments:

Either revision shows the power of time-order words. Use them appropriately, and you will be able to write with clarity and to communicate your ideas with readers smoothly.

4.2.10.3 Transition between paragraphs

Transition between paragraphs are accomplished with the same techniques used between sentences, except that the transitional elements may be different and longer. The transitional elements can be one of the following.

1) A sentence summarizing the preceding paragraph.
2) A question, at the end of the preceding paragraph, to be answered at the beginning of the new paragraph.
3) A transitional paragraph.

4.3 Conciseness

Clarity is essential and conciseness is important to writing. You can achieve conciseness by removing unnecessary extra words, phrases, clauses, sentences, and paragraphs. The removals should not sacrifice clarity, coherence, or unity.

Avoid overdose of traits, otherwise the writing becomes stuffy and wordy. In addition, avoid overuse of inappropriate jargons, stacked modifiers, and vague words.

Example 4-20. Using precise words for clarity

Vague: While traditional air-cleaning technologies such as these are prevalent, there are some creative solutions to air pollution being invested into.

Clear: While traditional air-cleaning technologies such as wet scrubbing and SCR are widely used, there are some novel solutions to air pollution being invested into.

Don'ts for Conciseness

- **Don't duplicate nouns of the same meaning.**

  Avoid the incorrect nouns in Table 4-3. You can use either the correct one or the acceptable one; they have the same meanings.

Table 4-3. Examples of duplicated nouns

Incorrect	Correct	Acceptable
Arrangement plan	Arrangement	Plan
Research work	Research	Work
Application results	Results	Application
Calculation results	Results	Calculation
Simulation results	Results	Simulation
Knowledge memory	Knowledge	Memory
Output performance	Output	Performance
Sketch drawing	Sketch	Drawing

- **Don't use redundant modifiers.**
  There are many redundant modifiers in literature. Table 4-4 shows typical redundant words relevant to engineering publications (Alred *et al.* 2018). *See also* Example 4-21.
- **Don't use a long, indirect expression.**
- **Don't use unneeded words or phrases inserted into a sentence.** *See* Example 4-22.

Table 4-4. Redundant words and phrases

Redundant	Concise
basic essentials	essentials
completely finished	finished
final outcome	outcome
present status	status
each and every	each, every
basic and fundamental	fundamental
first and foremost	foremost
perfectly clear	clear
completely accurate	accurate
with a view to	to
due to the fact that	because
for the reason that	
owing to the fact that	
the reason for	
by means of	by, with
by using	
through the usage of	
at this time	now, currently
at this point in time	
at the present	

- **Don't overuse or misuse of intensifiers.** *See* Example 4-23. Avoid overuse or misuse of intensifiers such as *best*, *great*, *more*, *most*, *very*, and the like in engineering scholarly writing.

Example 4-21. Redundant emphasis

Redundant: At its core, climate change is caused primarily through pollution, specifically the greenhouse effect. *[Source: Unpublished project report by a native English speaker]*.

Correct: ~~At its core, c~~Climate change is caused primarily through pollution, specifically the greenhouse effect.

Example 4-22. Unneeded words or phrases

Wordy: There are many researchers who have reported the effects of air pollution on the environment and public health.

Concise: Many researchers have reported the effects of air pollution on the environment and public health.

Wordy: Thousands of professionals will be attending the annual conference, which is scheduled on the 25th of next June.

Concise: Thousands of professionals will be attending the annual conference scheduled on the 25th of next June.

Example 4-23. Misuse of intensifiers

Incorrect: The experimental results and the model agreed very well.

Correct: The error between the experimental and model results is 5% of less.

Another commonly misused word is *significant*. *Significant* is used in a conclusion supported by statistical analyses. (*See* Example 4-24.) Without an indication of significance quantified by statistics, you need to present only the data without imposing your opinion on the readers. This type of ambiguous words should be replaced with specific and useful details for the readers to make the judgment.

Example 4-24. Use word *significant* in statistics

Linear regression analysis was used to indicate correlations, and any correlation with an R^2 value of greater than 0.8 was defined as statistically significant.

4.4 Parallel structures

Parallel structures achieve *emphasis, clarity,* and *conciseness* with less words, phrases, or clauses. Parallel structures also allow readers to anticipate the meaning of elements in the sentences (or paragraphs).

An effective parallel structure can be constructed by repeating an element of the sentence. The elements can be articles, pronouns, prepositions, and so forth. The sentence that you just read is in fact parallel structured. A parallel structure that begins with words, phrases or clauses must keep on with words, phrases or clauses. (*See also* 5.1; 6.1; Example 4-25.)

Example 4-25. Parallel structures in writing

Parallel words:

Corona viruses carry either DNA or RNA, never both.

A sophisticated air quality monitor may be used to detect these air pollutants: sulfur dioxide, volatile organic compounds, ozone, and particulate matter.

Parallel phrases:

Both the medical professionals and the patients are exposed to the airborne viruses.

Parallel clauses:

Either we work together to fight the pandemic, or we let the pandemic beat us all.

Your technical proposal not only must have a list of objectives, but also should include a list of deliverables.

In typical scholarly writing, the authors are expected that they would start with background information, that they would elaborate on the methodology, and that they would present the results to the readers.

4.5 Writing with Professional Style

To summarize briefly, professional writing for engineering publication has its own style beyond an individual's personal and distinct stylistic traits. This applies to engineering and scientific articles (prepared for academic journals or proceedings of international conferences), books, encyclopedia, Master's theses, and doctoral dissertations. Therefore, the writer's tone should be impersonal and objective, rather than subjective, for the greater benefit of society instead of the writer's personal opinions.

Academic writing is expected to be clear and direct. The vocabulary ought to be specialized and precise instead of fancy, elegant, or dull. Therefore, contractions, slangs, or dialects are not frequently used in the academic writing. The key is to keep your readers in mind; they are from diverse cultural background with different levels of education and experience. Ask yourself this question while you are writing: what is important to the readers?

On the other hand, professional writing can be interesting and lively too. Monotonous style can be avoided by using active voice, positive thoughts (Example 4-26), various structures, and balanced emphasis and subordination. However, intensifiers (*most*, *much*, *very*) should be used for emphasis with caution.

Example 4-26. Positive and negative thoughts

Negative: Results show that 70% of the conclusions are misleading.

Positive: Results show that 30% of the conclusions are supported with data.

In addition, communicate with goodwill and modesty (Example 4-27). For a publication aimed at international readers, who may be foreign to the subject, you would rather provide more details than omitting too much. Modesty is more acceptable than arrogance.

Example 4-27. Writing with goodwill and modesty

Arrogant: Our device can capture particles at a 100% of efficiency.

Modest: Based on the results herein, we can say that our device can capture particles at an efficiency of almost 100%.

These basic principles and high-level techniques should get you ready for drafting. Now, you can try the basic principles in your writing. The rest of this chapter contains the guidelines about the first draft of a typical document, from the title to the appendices, explained with examples.

4.6 Draft Title

The titles of documents are important to both the writers and the readers. An effective title condenses the scope and purpose of the writing into one concise clause. It is expected to call for readers' attention and interest. The titles should be concise and accurate with a proper length. The following "don'ts" should help you determine the titles with clarity and conciseness.

Don'ts in titles

- Don't use words *On*, *Studies on*, *A Report on*, or the like. [*See* Example 4-28 (Siddiqi *et al.*, 2001)]
- Don't use abbreviation, including chemical formula.
- Don't use the sentence form or a rhetorical question.

Example 4-28. Redundancy in the title of a journal article

Don't A Study of the effect of nitrogen dioxide on the absorption of sulfur dioxide in wet flue gas cleaning processes.

Do: ~~A Study of the~~ Effect of nitrogen dioxide on the absorption of sulfur dioxide in wet flue gas cleaning processes.

However, they are acceptable to the *subtitles* because they specify the purposes and scopes of the reports. More can be included in the subtitles – that is what they are for. For example, United States Environmental Protection Agency's (US EPA) publishes annual reports on new light-duty vehicle greenhouse gas (GHG) emissions, fuel economy, technology data, and auto manufacturers' performance in meeting the agency's GHG emissions standards. The subtitle contains much information than the title itself. The annual report published for 2019 has a title of *The 2019 EPA Automotive Trends Report* with a subtitle of *Greenhouse Gas Emissions, Fuel Economy, and Technology since 1975*. (EPA, 2020)

Example 4-29. Book titles: subtitles and article titles

Book titles:

- Nanoengineering: Global Approaches to Health and Safety Issues.
- Handbook of Polymer Nanocomposites for Industrial Applications.
- Cognitive Informatics, Computer Modelling, and Cognitive Science: Volume 1: Theory, Case Studies, and Applications.

Article titles:

- Selection and interpretation of diagnostic tests and procedures.
- Coupling and entangling of quantum states in quantum dot molecules.
- Efficient bipedal robots based on passive-dynamic walkers.

4.7 Authorship and Affiliations

4.7.1 Authorship

Authorship has become an unnecessarily complicated issue with the growing demand for multiple disciplinary research collaboration. In some countries, only first author or the contact author(s) merit career advancement, while in others, all co-authors are considered equally important to the work. The number of publications may impact their career advancement (such as tenure and promotion) or personal incomes (*e.g.* grants and bonuses). There is no single definition of *authorship* that applies to the world. It is up to the co-authors to withhold their academic integrity and responsibility of the published works.

It is ethical to add co-authors who contributed to the merit of the work. Most journal articles, technical reports, and similar works are written by teams with complementary expertise, whereas degree theses and dissertations are most likely by individual students who received the degrees. Regardless of the publications, the co-authors should be the writers or the investigators or the team members who contributed to the knowledge production. The contributions can be active involvement or supervision of one or more of the following activities:

- Concept design, methodology development, data collection, or data analysis.
- Manuscript drafting or revisions.
- Review and approval of final version for submission or publication.
- Ensuring accuracy and integrity of the research and publication.

Listing too many co-authors dilutes the contributions of each (Example 4-30). It may look unreasonable, especially to the fellow professionals, who know how much efforts it takes to produce certain type of engineering works.

Example 4-30. Authorship with diluted contributions

Spatial and seasonal distributions of carbonaceous aerosols over China

J. J. Cao, S. C. Lee, J. C. Chow, J. G. Watson, K. F. Ho, R. J. Zhang, Z. D. Jin, Z. X. Shen, G. C. Chen, Y. M. Kang, S. C. Zou, L. Z. Zhang, S. H. Qi, M. H. Dai, Y. Cheng, K. Hu

4.7.2 Authorship order

Consult with your superior before finalizing the order of authorship. It may be different from what you have learned before. In many disciplines and countries, the authorship order indicates the weight of contribution. Most agree that the first author contributes most to the work and that the last author representing predominantly the one who overseas the entire work. However, this does not apply to all countries. In many countries, for example, the most senior may be listed as the second author and the corresponding author.

Example 4-31 shows the title and authorship of the article written by Fuller *et al.* (2012). I requested through the Copyright Clearance Center and received the permission from the American Chemical Society (ACS) to use a large body of the article as examples in this book. The sample paper was written by native English speakers. More examples from the same reference are used as well-written examples throughout this book.

Example 4-31. Title and authors of the sample paper

Direct Surface Analysis of Time-Resolved Aerosol Impactor Samples with Ultrahigh-Resolution Mass Spectrometry

Stephen J. Fuller† Yongjing Zhao‡ Steven S. Cliff‡ Anthony S. Wexler‡ Markus Kalberer*†

† Department of Chemistry, University of Cambridge, Lensfield Road, Cambridge CB2 1EW, U.K.
‡ Air Quality Research Center, University of California−Davis, Davis, California 95616, United States
E-mail: markus.kalberer@atm.ch.cam.ac.uk. Phone: +44 1223 336392.

The corresponding author is typically the one who oversees the work, and who has a permanent position where the work was produced. The corresponding authors of works produced at universities are normally the academic supervisors of the research teams instead of the students. The students often leave the universities after completing the research works. Additionally, students may not have the depth of knowledge to respond to readers' inquiries.

4.7.3 Affiliations

Most publishers require affiliations to be listed right after the names of co-authors. Make sure that the affiliation is the institution where the co-author made the contributions to the work instead of the paper – the paper and the work are different concepts. If a university student, for example, is employed by a company after successful submission of the dissertation, the university should be listed as the affiliation instead of the current employer when a journal article is written based on the student's thesis work.

Example 4-32 shows the title, authorship, and affiliations of an article that was published by my former graduate student and a collaborator in another country (Givehchi *et al.* 2018). The first author (Givehchi) carried out the research work at the University of Waterloo under the joint supervision of the second author (Li) and the last author (Tan). They all have major contributions to the research work and the writing of the article. At the time of publication, Givehchi was employed by the University of Toronto. However, the University of Toronto cannot be listed as Givehchi's affiliation in the published article. Z. Tan was affiliated with both institutions. Q. Li and Z. Tan are listed as contact authors because either one is qualified to answer any questions arising from the article.

Example 4-32. Title, authorship and affiliations

Filtration of Sub-3.3 nm Tungsten Oxide Particles Using Nanofibrous Filters

Raheleh Givehchi[1], Qinghai Li[2,*] and Zhongchao Tan [1,2,*]

[1]Department of Mechanical & Mechatronics Engineering, University of Waterloo, Waterloo, ON N2L 3G1, Canada; raheleh.givehchi@utoronto.ca

[2]Tsinghua University—University of Waterloo Joint Research Centre for Micro/Nano Energy & Environmental Technologies, Tsinghua University, Beijing 100084, China.

*Correspondence: liqh@tsinghua.edu.cn (Q.L.); tanz@uwaterloo.ca (Z.T.)

4.8 Drafting Abstract

The abstract of your publication allows the readers to decide whether it is necessary to continue reading beyond that point. The abstract should be treated as a stand-alone document and it needs to tell a complete story apart from the original document.

The length of an abstract varies with the document and organization. It is typically 200 to 250 words for an article and about 1000 for a thesis. Extended abstracts, normally 3-5 pages, are collected in proceedings of international conferences. Regardless of their lengths, all abstracts follow similar principles.

There are two types of abstracts, *descriptive abstract* and *informative abstract.* The former summarizes the objectives and methodology in the research; it is relatively short. The latter can be considered as an expansion of the descriptive abstract with additional information such as results, conclusions, and recommendations. You need to use descriptive abstracts in compiled documents, such as a collection of conference papers and technical reports of a large project before the final one is ready. Informative abstracts work best for a wide range of readers who are likely interested in the conclusions and recommendations.

A well-written abstract follows the *Why-How-What* pattern. It contains three key elements: objective (*Why*), methodology (*How*), and results or conclusions (*What*). A short abstract with word limit should begin with the subject and the scope of the work. However, many writers, including native English writers, often precede the topic sentence (objectives) with background information. As a result, the abstract does not have enough information on results or conclusions. Example 4-33 shows a well-written abstract (despite its imperfection) that follows the sample paper (Fuller *et al.* 2012).

Example 4-33. A well-structured abstract

Aerosol particles in the atmosphere strongly influence the Earth's climate and human health, but the quantification of their effects is highly uncertain. The complex and variable composition of atmospheric particles is a main reason for this uncertainty. About half of the particle mass is organic material, which is very poorly characterized on a molecular level, and therefore it is challenging to identify sources and atmospheric transformation processes. We present here a new combination of techniques for highly time-resolved aerosol sampling using a rotating drum impactor (RDI) and organic chemical analysis using direct liquid extraction surface analysis (LESA) combined with ultrahigh-resolution mass spectrometry. This minimizes sample preparation time and potential artifacts during sample workup compared to conventional off-line filter or impactor sampling. Due to the high time resolution of about 2.5 h intensity correlations of compounds detected in the high-resolution mass spectra were used to identify groups of compounds with likely common sources or atmospheric history.

Comments:

The is a well-structured abstract following the *Why-How-What* pattern. The abstract begins with the *Why* [the research is important and necessary]: uncertainties exist, aerosol particles are poorly characterized, *etc.* Then it introduces *How* [to address the knowledge gaps]: a new technique. Finally, the abstract ends with *What* [are the findings]: groups of compounds with common sources.

However, this abstract could be further refined. It could have begun with a shorter *Why* and ended with a longer *What.* The *Why* part has 67 words; the *What* part has less than 34 words and it lacks details, although the authors were allowed to add 45 more words into the abstract. [Note: *Abstracts in Analytical Chemistry should have 80-200 words.*]

Writing an excellent abstract is challenging, even to the native English speakers. Therefore, it is normal for the non-native English speakers to make some mistakes in writing the abstract. Example 4-34 and Example 4-35 are abstracts written by non-native speakers, and they illustrate some of the typical mistakes that non-native English writers may make in writing abstracts.

Example 4-34. A poorly-written abstract

The particle number size distribution (3 nm-10 μm) was conducted in May of 2011 at the urban sampling site of x City. Their pollution characterization and their dependencies on gaseous contaminants and meteorological factors were synchronously investigated. The diurnal average total of particle number were 568 cm^{-3} in nucleation mode (3-20 nm), 14,909 cm^{-3} in Aitken mode (20-100 nm), 7418 cm^{-3} in accumulation mode (0.1-1 μm) and 2 cm^{-3} in coarse mode (1-10 μm), respectively. The particle number, surface area and volume size distributions respectively presented the single peak, double peak and four peaks pattern. The diurnal variation of particle number in nucleation mode was mainly influenced by nucleation events in spring. The diurnal variation of particle number in Aitken mode was closely correlated with the traffic densities. The maximum contribution ratios to the total of the particle number, surface area and volume concentration were particles in Aitken mode, accumulation mode and accumulation mode, respectively. ......The absolute values of correlation coefficient between other factors (SO_2, NO_2, PM_{10}, PM_1 and visibility) and the particle number concentration was maximum in accumulation mode and indicating that the most notable method was decreasing the particle number concentration in accumulation mode for improving atmospheric environment quality. Local wind speed played an important role in shaping the particle number size distribution in the urban area of Jinan, China. With the increasing of wind speed, the particle number concentration increased in nucleation mode and decreased in the Aitken mode and accumulation mode. The particle number concentration in haze weather was evidently higher than that in clear weather. The union action of gaseous contaminants and meteorological factors was supposed to supply a favorable environment for the formation and growth of secondary particles to lead the lower visibility in haze weather.

Comments:

This abstract has several key structural and language errors. First, it begins with one sentence for *Why* (the scope), followed by another sentence for *How* (the method). Both *Why* and *How* lack clarity. The *What* part of the abstract is then flooded with information at a rapid pace. This 293-word long abstract should exceed the word limits of most journals. Finally, this wordy abstract lacks both conciseness and clarity. It becomes useless and boring to many readers.

Abstracts are expected to be written with clarity and conciseness. Do not omit articles or important transitional words and phrases. You can use acronyms after definition, but avoid citations, equations, references, and the like. Example 4-35 shows an abstract published in an international journal. It has these types of errors (Du *et al.* 2006).

Example 4-35. A poorly-written abstract with errors

In the computational fluid dynamics (CFD) modeling of gas–solids two-phase flows, drag force is the only accelerating force acting on particles and thus plays an important role in coupling two phases. To understand the influence of drag models on the CFD modeling of spouted beds, several widely used drag models available in literature were reviewed and the resulting hydrodynamics by incorporating some of them into the CFD simulations of spouted beds were compared. The results obtained by the different drag models were verified using experimental data of He et al. [He, Y.L., Lim, C.J., Grace, J.R., Zhu, J.X., Qin, S.Z., 1994a. Measurements of voidage profiles in spouted beds. Canadian Journal of Chemical Engineering 72 (4), 229–234; He, Y.L., Qin, S.Z., Lim, C.J., Grace, J.R., 1994b. Particle velocity profiles and solid flow patterns in spouted beds. Canadian Journal of Chemical Engineering 72 (8), 561–568.] The quantitative analyses showed that the different drag models led to significant differences in dense phase simulations. Among the different drag models discussed, the Gidaspow (1994. Multiphase Flow and Fluidization, Academic Press, San Diego.) model gave the best agreement with experimental observation both qualitatively and quantitatively. The present investigation showed that drag models had critical and subtle impacts on the CFD predictions of dense gas–solids two-phase systems such as encountered in spouted beds.

Comments:

This abstract is well-structured but full of writing errors. The most obvious blunder is the presence of references, one enclosed by brackets and another by parentheses. There are 220 words in this abstract, and 68 words are wasted on references. Furthermore, the authors misused intensifiers such as *significant, best, critical,* and *subtle,* which are vague words. (*See also* 6.1.2 and 7.2.2)

4.9 Drafting Executive Summary

Executive Summary is normally considered as part the body of a long document. It is introduced here, however, because an executive summary has similar functions as an abstract. The main body of a long document, such as a report, begins with an executive summary, which summarizes the highlights from the full document. An executive summary may be published separately representing the original document. Therefore, do not cross-reference to contents in the main document.

A well-written executive summary is a condensed version of the entire document; it should also be written with accuracy, clarity, and conciseness. It may contain objectives, scope, methods, results, conclusions, and recommendations, if applicable. To maintain its independence from the original document, the executive summary may have essential figures, tables, or footnotes.

Executive summaries are usually about 10 % of the length of the main documents, with key divisions following similar order of sequence. For example, the Executive Summary of *The (US) EPA Automotive Trends Report - Greenhouse Gas Emissions, Fuel Economy, and Technology since 1975*" is 13 pages long, and the full report has 155 pages (EPA, 2020).

4.10 Keywords

Keywords are usually listed right after the abstract or the executive summary. Keywords are used for indexing by database and search engines to facilitate easy discovery of the original document. With the advances in online access to knowledge, most documents are now published online. Electronic search engines, databases, or journal websites use the keywords to identify whether and when to display the documents to potential readers. Keywords, title, and abstract are three key elements that can improve the discoverability of your work. Without them, your work may only reach a small portion of the potential readers in the world.

Keywords should be selected after careful consideration. To choose the right keywords, you can start with a list of terms and phrases that are used repeatedly in your document. Make sure that your list of keywords includes all the key phrases, terms for procedures, common abbreviations *etc.* Then type your keywords into a search engine and check if the results that show up match your keywords. This may not be the perfect solution, but it helps you determine whether the keywords in your publication are appropriate.

4.11 Table of Contents

Tables of contents (TOCs) are needed in long documents, excluding most journal articles and conference papers. A TOC enables quick reviews of the document organization because the TOC lists all the major headings and their page numbers. A typical TOC contains levels one to three (or one to two) headings. A TOC with more than four levels of headings would create visual clutter.

4.12 List of Figures

The List of Figures identifies the titles and locations (page numbers) of the figures (charts, drawings, photographs) in a long document. Articles in periodicals and proceedings of conferences do not normally use lists of figures.

4.13 List of Tables

Like the function of List of Figures, the List of Tables identifies the titles and locations (page numbers) of the tables in long documents. Articles in periodicals and proceedings of conferences do not normally use lists of tables. The List of Tables allows readers to quickly and easily locate the tables of interest.

Note:

Only long documents need Table of Contents, List of Figures, and List of Tables. They can be created automatically using word processing software.

4.14 List of Abbreviations, Acronyms and Symbols

Abbreviations and acronyms are frequently used in technical writing and engineering publications. An abbreviation is a shortened word, for example, *Dr.* for *doctor*. An acronym is a word created using the first letters of the words in a phrase, such as *DOE* (Department of Energy) and *UW* (University of Waterloo). The proper use of abbreviations and acronyms enhances the readability and efficient comprehension of the document they belong to.

Most abbreviations can be used without definition in the document because they are accepted by the professional in the field, and occasionally the public. For example,

- Social titles (*Mrs.*, *Mr.*) and professional titles (*Dr.*)
- Directions in city address: *SW* for south west.
- International units: *mm* for millimeter; *s* for second; *kg* for kilogram; *K* for Kelvin; *Pa* for Pascal.

Units of measurement are used in their abbreviated forms. The International System of Units (abbreviated SI) should be used for publications aimed at international readers (*see* 9.9). Dual systems of units are occasionally used in English documents, especially those traditionally published in the United States of America.

Avoid an acronym without definition. Define an acronym right after its first appearance in the text (*see* Example 4-36). For long documents (*e.g.* thesis, books), you may redefine the same acronyms for different chapters. It should help the readers who read only part of the long documents.

Example 4-36. Defining and using acronyms

Research in Geographic Information Systems (GIS) is advancing rapidly… GIS is an increasingly important technology to the military.

For occasionally used abbreviations or acronyms in a long document, reverse the definition (with the full text in parentheses) if it appears at a reasonable interval. *See* Example 4-37. It helps remind readers of its meaning because they may have forgotten how it is defined when it first appears.

Example 4-37. Reverse definition of acronym

He is an expert in GIS (Geographic Information Systems)

Acronyms usually act as nouns in sentences. When an acronym follows an article (*a* or *an*) in a sentence, choose the right article based on the sound rather than the original phrase of the acronym. For example, *an unmanned aerial vehicle* (*UAV*), but *a UAV*. Acronyms can generally be pluralized with the addition of a lowercase -s (*e.g.*, *UAVs*). The acronyms can also be made possessive with an apostrophe followed by a lowercase *-s* (*UAV's*).

Many professional organizations define their own acronyms. Learn and use their acronyms correctly. For example, *ASME* is for *American Society of Mechanical Engineering*, but the *American Institute of Chemical Engineers* uses the acronym *AIChE* instead of *AICE*. As another example, *PE* and *PEng* are used for the same text of *Professional Engineer* in the United States of America (USA) and Canada, respectively.

Acronyms are subjects. They should be treated as singulars, which require singular verbs (*e.g.,* AIChE was established in 1908), even though they stand for plurals.

Don'ts for Abbreviations:

- Don't create your own abbreviations; instead, use widely accepted ones.
- Don't use periods in uppercase acronyms (NDA, IRA); use period marks for lowercase acronyms (a.m., *e.g., etc.*).
- Don't follow an abbreviation or an acronym with double period marks.

4.15 Foreword

The *Forward* is an optional introduction of a long document, such as a book or a formal report. The Forward is normally written by an influential and important person with authority in the field. The foreword introduces the author (s) and the document to the readers. A nicely crafted foreword can also serve as an endorsement for the value of the document.

Notes:

The *Forward*, like a letter to the readers, is signed and dated at the end. The *Foreword* precedes the *Preface* when both exist.

4.16 Preface

The *Preface* is another optional introduction of a long document. It is written by the author(s) of the original document. Most books have Prefaces. A preface introduces the contexts, including driving force behind the writing, the motivation of the work, and the purpose of the document. A typical preface also contains acknowledgments to people who have contributed to the preparation and publication of the work. (*Acknowledgments* can be a stand-alone section located at the near end of a short document such as a journal article.)

4.17 Drafting Introductions and Objectives

The *Introduction* provides readers with enough background information about the work. The background information may be a state-of-the-art review of earlier works in the field, serving as a context, leading to a statement of the objectives of your work. In scholarly articles aimed at advancing knowledge and technology, the statement of objectives clarifies an existing perspective and the knowledge gaps in the field. No one single piece of work can fill up all the knowledge gaps, and the scope of the document you are preparing should be specific and direct for the readers. It is also necessary to describe how you plan to develop the subjects of long documents like multi-chapter theses. This information allows readers to preview the structure of your documents, and even helps the readers to evaluate your methodology, conclusions or recommendations.

4.17.1 Sources of information

The *Introduction* or *Literature Review* aims at a better understanding of the state of the art. The background information collected and reviewed are from various sources, including online or printed documents, for comprehensiveness. Some researchers may also consult internal archives for the most relevant information. While peer reviewed articles and books are considered creditable sources of literature, peer reviewed documents published on academic and government websites are normally acceptable. Library is recommended for locations where online resources are not accessible.

You ought to execute due diligence to ensure the validity of online sources. In some countries, online resources are abundant and easy to access, but it is not always the case in many other countries. In an open society, internet provides access to most of the information that we need. However, it is challenging to verify the completeness and accuracy of some information because anyone can publish online. A safe approach to valid information is to access creditable sources. The online versions of articles in reputable journals, books by prominent authors, and the like would have the same merits as the printed versions.

Regardless of the scope, keep in mind that the more state-of-the-art the information, the better. Using only outdated information generally undervalues your work because there is a high chance that you miss recent advances in the same scope of research. Lacking state-of the-art information may indicate the uselessness of your work. It might also be viewed as incapability or dishonesty of the author(s). Again, writing reveals who you are.

The *Introduction* of a short documents may be merged with *Literature Review*, which is introduced as follows.

4.17.2 Literature review

Literature Review varies in length, depending on its purpose. It can be one section in a short document or one chapter in a thesis or dissertation. A comprehensive literature review, as a key element of the *Introduction*, establishes the importance and ensures the novelty of your work. *Literature review* may also become a stand-alone article for the synthesis of knowledge.

Literature review is more than a compilation of previously published documents. As you review each source, evaluate its validity, investigate its limitations, provide your criticism, and judge its value to the readers. You can arrange your evaluation and criticism chronologically or in various subcategories of the topic.

Figure 4-2 illustrates the **funnel approach** to literature review. A thorough literature review begins with a relatively big picture, narrows gradually to the state of the art, funnels naturally to specific knowledge gaps, and ends at focused objectives. At the end, you articulate the questions to be answered, the problems to be solved, the technologies to be developed, the methodology to be improved, and so on. Finally, a transition is needed to naturally lead your writing to the next part of the document, *Objectives*.

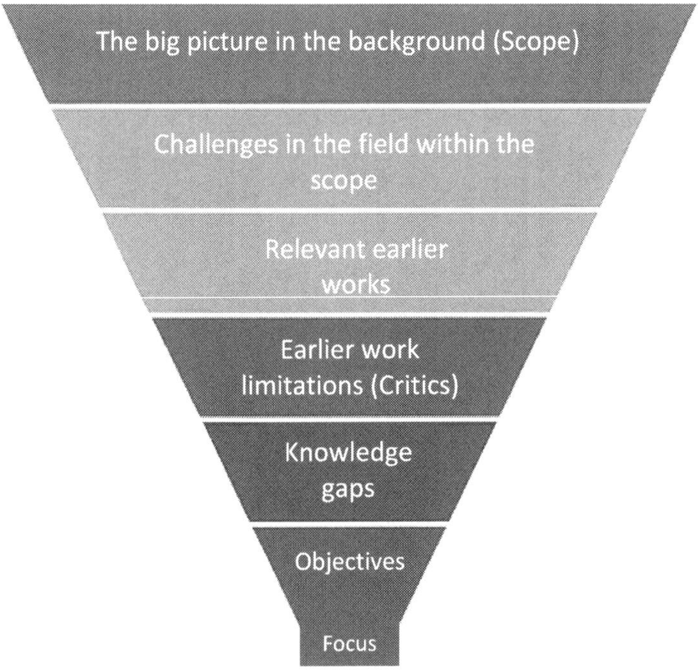

Figure 4-2. Funnel approach to literature review

The funnel approach to literature review is critical to the justification of the publication of your work. It is also important to the clarity of your writing. Therefore, this useful technique is further explained with Example 4-43 through Example 4-49 (Fuller *et al.*, 2012).

Example 4-43. The first paragraph of the Introduction

Aerosol particles are important components of the atmosphere and strongly influence the Earth's climate by directly scattering or absorbing sunlight and indirectly by aerosol−cloud interactions. Aerosolparticles are also involved in negative health effects caused by air pollution and are linked to increases in respiratory and cardiovascular diseases. All these particle effects are influenced by the chemical composition of the aerosol particles. A major fraction, often more than 50% in mass, oftropospheric aerosol is organic material, which isvery poorly understood on a molecular level, although several thousand compounds have been separated with chromatographic techniques. To identify sources of particles, but also to understand compositional changes during their atmospheric lifetime, an automated method with high time resolution is highly desirable.

Comments:

The *Introduction* begins with the *importance* of the work (*atmosphere, climate, health, etc.*). Keeping in mind that they are not the scope of this work, the authors quickly narrow it down to *chemical composition of aerosol particles* and highlight the limitations in the field (*organic materials, poorly understood*). At the end, they bring up the need of the work (*automated method, high resolution, highly desirable*).

Most journals require elaboration on this big picture. Usually the first one to two paragraphs give a high-level introduction of the background information. However, this prominent journal does not give the authors much room for the elaboration on the big picture.

Each of the sentences in this paragraph can be extended into a paragraph of a long document.

Example 4-44. The second paragraph of the Introduction

The analysis of the organic fraction of atmospheric aerosol particles on a molecular level is often challenging because their composition can be highly variable in time and space and usually only small sample amounts are available for analysis because atmospheric concentrations are typically a few micrograms per cubic meter. The composition also depends on the size fraction of the aerosol; for example, resuspended, wind-blown particles are mostly larger than 1μm and have often a very distinct chemical compositioncompared to particles smaller than 1μm, which are mainly emitted by combustion sources or formed by chemical reactions in the atmosphere.

Comment:

Following the first paragraph, the ideas in the second paragraph become more specific by identifying the *challenges* and the general *scope* of work.

Example 4-45. The third paragraph of Introduction

Thus, analytical−chemical techniques used to analyze atmospheric organic aerosol particles need to be highly sensitive to allow for highly time-resolved analyses at trace concentrations. In addition, the analysis technique needs to be able to characterize highly complex organic compound mixtures with thousands of mostly unknown components.

Comments:

The idea is further narrowed down to highly sensitive analysis of organic aerosol particles. The authors seem to set the stage for the major contributions to be introduced in *Conclusions*. However, this short paragraph could be better structured. (*See* 5.1. Paragraph Structures.)

Example 4-46. The fourth paragraph of Introduction

There are a number of analytical−chemical techniques that allow measuring the composition of aerosol particles with high time resolution. Online aerosol mass spectrometry (AMS) techniques, for example, allow for very highly time-resolved particle composition studies. However, such measurements are demanding with respect to manpower and other resources and are usually performed only for a few weeks at a specific site. Thus, such measurements rarely provide insight into long-term trends of particle composition. In contrast, aerosol samples collected on filters or impactors over long time periods are more readily available but have mostly a rather low time resolution of the order of a day or more, and their chemical analysis usually involves time-consuming.

Comments:

This paragraph summarizes earlier works related to the scope of the current study. The authors provide criticism (*however*, *but*) and evaluate the limitations of the state of the art. They imply the novelty of the current study.

Nonetheless, a literature review is not a simple compilation of the earlier works. It is important to comment on the state of the art by identifying their limitations, which are also new research opportunities. The limitations should be related to your own study, which addresses the knowledge gaps between earlier studies and the current one. Example 4-47, using the same paragraph in Example 4-46, illustrates how to identify the knowledge gaps (*i.e.,* the research opportunities).

Example 4-47. Identifying knowledge gaps

[*This is the same paragraph as in* Example 4-46] There are a number of analytical−chemical techniques that allow measuring the composition of aerosol particles with high time resolution. Online aerosol mass spectrometry (AMS) techniques, for example, allow for very highly time-resolved particle composition studies. <u>However, such measurements are demanding with respect to manpower and other resources and are usually performed only for a few weeks at a specific site.</u> Thus, such measurements rarely provide insight into long-term trends of particle composition. In contrast, aerosol samples collected on filters or impactors over long time periods are more readily available, <u>but have mostly a rather low time resolution of the order of a day or more, and their chemical analysis usually involves time-consuming workup and is prone to artifacts during sample preparation.</u>

Comments:

The underlined sentences identify the limitations of the earlier works; they lead to the knowledge gaps. The limitations of earlier studies are usually introduced by conjunctive adverb like *however*, *but*, and so on. (*See* Table 6-1. Conjunctions)

There must be other researchers who attempted to address these knowledge gaps. Therefore, the next paragraph reviews the state of the art that is *closely related* to these gaps.

Example 4-48. Reviewing closely related earlier works

[Fifth paragraph] The rotating drum impactor (RDI) is a sampling technique that allows for off-line chemical analysis of particles with a time resolution on the hour time scale, which is sufficient to resolve most atmospheric aerosol processes. The RDI has been combined with synchrotron X-ray fluorescence (s-XRF) analysis to measure changes in metal content directly from the RDI samples. Only one study analyzed so far organic components collected with an RDI: Emmenegger *et al.* investigated the temporal variability of polycyclic aromatic hydrocarbons collected at an urban location with two-step laser mass spectrometry, where an IR laser was used to desorb analytes directly from the RDI stripes without further sample preparation.

[Sixth paragraph] The recent development of commercially available surface mass spectrometry ionization techniques such as liquid extraction surface analysis (LESA) and desorption electrospray ionization (DESI) enables analysis of a wide range of organic compounds with high spatial and therefore high time resolution from RDI samples (for details see the Experimental Section). LESA has been used previously for the analysis of biological samples and pesticides, but applications for environmental samples have not yet been described in the literature. Another online extraction technique, nano-DESI, was recently presented by Roach et al. and was applied to atmospheric aerosol filter samples. Importantly, LESA and (nano-) DESI require no off-line sample preparation, such as solvent extraction and solvent evaporation. Reducing the number of sample preparation steps also reduces the possibility of introducing artifacts.

Comments:

Following the fourth paragraph, the fifth and the sixth paragraphs in the sample paper (Fuller *et al.* 2012) identify and evaluate the most recent and the most relevant earlier works that seem to address the knowledge gaps. This step precedes the Objectives.

4.17.3 Objectives

The state-of-the-art literature review eventually funnels down to the *objectives*. The objectives must be specific and exact questions to answer; they must be focused and closely related to the scope of your own study as indicated by the *Title*. You can pick some scientific questions to answer through your work for the unity of writing. You should separate them into different short documents (articles) or different sections (chapters) of a long document if you have multiple irrelevant objectives in mind. Again, funnel down to coherent objectives. Don't suddenly stop your literature review. Example 4-49 shows how Fuller *at al.* (2012) funnel their preceding review down to their objectives.

Example 4-49. Funneling down to the objectives

For this study RDI collection was combined for the first time with LESA and ultrahigh-resolution mass spectrometry (UHRMS) to characterize organic components in aerosol particles and to observe changes in ambient concentration with a time resolution of about 2.5 h for sub- and super-micrometer particles. UHR-MS allows identification the elemental composition of unknown compounds and thus is a very valuable technique for the analysis of highly complex and often poorly characterized atmospheric organic aerosols where often the majority of the compounds are unknown.

Comments:

I see two objectives in this paragraph. The first one is to develop a new technique (*…for the first time*), and the second, to characterize *the organic compounds in aerosol particles…*

4.17.4 Typical introduction and objective errors

Example 4-43 through Example 4-49 illustrate the well-written *Introduction* and how it funnels down to the *Objectives.* It is more challenging to many writers than it appears. All writers make mistakes in writing. With continual learning and practice, however, you will make less and less mistakes. Beginners and intermediate-level writers often make the following writing mistakes, which are typical to non-native English speakers.

- **Unclear "funnel" shape**
  The "funnel shape" is critical to the effectiveness of the introduction, although this structure is not always necessary to writing in another culture or language.
  A broken funnel may result from one or more of the following errors.

a. Insufficient breadth of background, at the top of the funnel.
b. Insufficient explanation of importance, in the middle of the funnel.
c. Insufficient justification of the novelty, near the bottom of the funnel.
d. Insufficient summary of prior arts.
e. Unclear knowledge gaps.
f. Multiple directions leading to irrelevant questions to be answered.
g. Incomplete literature without clear goals, when the literature review stops suddenly.

- **"All-mighty" objectives without focus**. (*See* Example 4-50)

Example 4-50. Over-generalized objectives

……~~Unfortunately~~, by digging out in literature, we found that a detailed examination to the effect of drag models on CFD modeling seems unavailable. In spouted bed systems, the volume fraction of particles can vary from almost zero to the maximum packing limit, leading to much more complex behavior of drag forces than that in normal fluidization systems. By incorporating various drag models into the two-fluid model, the present study is conducted with the aim of fully understanding the influence of the choice of drag models on simulation and thereby laying a basis for the CFD modeling of spouted beds. (Du *et al.* 2006)

Comments:

It takes numerous studies to understand most of a discovery. Similarly, the basis for CFD modeling should have been established by other researchers too. The specific incremental contributions in one study should become the specific objective(s).

- **Confusing tasks with objectives**

New and intermediate writers often mistaken the tasks with the objectives. A task is part of the methodology, indicating the investigations needed to achieve a specific objective, but the task is not an objective. Example 4-51 shows the task-like objectives listed in a thesis.

Example 4-51. Tasks and objectives

The objectives and significance of this study are as follows:

Develop experimental apparatus to be used for obtaining kinetic data for CO_2 absorption into select solutions. The apparatus will be a continuously stirred tank reactor (CSTR) reactor equipped with gas and liquid analysis instruments.

Evaluation of the activity of the CA enzyme in the PC solution. This part of the research will quantify the effectiveness of the CA enzyme to promote the absorption rate of CO_2 into the PC solution and compare the promoted rate with the un- promoted rate and with the MEA system. The activity (kinetic) data obtained from this study are ~~significant~~ because they will provide the basic information required to evaluate the effectiveness of the CA enzyme and for design calculations, process optimization and cost analysis for the scale-up of the IVCAP.

......

Development of a mathematical model to simulate the absorption of CO_2 into the PC-CA solution in a stirred tank reactor. This model will be evaluated against experimental data. The significance of the evaluated model is that it will be used to predict the absorption flux at operating conditions beyond that of experimental tests (e.g. high temperatures, high enzyme concentrations). The model can also guide the future experimental design and the design calculations of the resulting absorption column.

Comments:

The topic sentences of all these paragraphs indicate the tasks to be completed over the graduate studies. The objectives behind them are not clearly stated. Readers have to guess. In the first paragraph, for example, to *develop experimental apparatus to be used for obtaining kinetic data* is a typical task in engineering research. So are true for *the evaluation of the activity of the CA enzyme* and *the development of a model.* Data and models are used for certain purposes, which should be clearly stated as objectives.

4.18 Drafting Methodology

Clear description of methodology is important to research training. Engineering publications are not necessarily for the experts. They are often the opposite as their readers include many undergraduate and graduate students, who are entering the fields. Clarity even becomes an ethics matter for description of experimental procedures that involve safety risks. This applies to most original engineering publications, including journal articles, books, and graduate theses.

The methodology for experimental studies that involve laboratory testing, or a major investigation, must contain enough details allowing readers to reproduce the works and validate the results if they wish to do so. Otherwise, the results are not considered validated or creditable. The details include the equipment and devices, the materials and supplies, and the apparatus and components used in the experiment.

In addition, you need to justify your choices of approach to the research by comparing with alternative approaches described in earlier publications. In addition, you should emphasize the novelty of your approach to the problem. All these factors matter to the accuracy of your results and the validity of your conclusions. You may include the observations and problems encountered, if any. They save your readers' time from repeating the same mistakes.

4.18.1 Numerical approaches

Either numerical or experimental approach is one of several approaches to solving a problem. Like experimental procedures, numerical modeling procedures should also include sufficient details. Clearly explain the physical models and simplifications made to solve the equations. In short, you need to clearly write down the following information.

- **How the approach is developed and used in your wok.**

Typical numerical models are developed from theoretical analyses using many equations. These equations describe the general principles in the specific field, such as conservations of mass, momentum and energy. They are usually simplified to match the conditions of the specific problems.

- **What specific assumptions or simplifications are made for the numerical models.**

The general principles cannot solve the specific problems. Most likely you need to make reasonable assumptions and simplifications before you can proceed with the numerical simulation. These are the factors that may affect the accuracy of results and conclusions, but the simplifications make your studies affordable and acceptable. It is a comprise between accuracy and cost. Therefore, you must clearly describe and justify the assumptions and simplifications. In addition, the boundary conditions and the initial conditions should also be described with details. Different simplifications and conditions may lead to different results and conclusions.

- **What platform you have used to solve the models**

You may use the latest version of commercial software to solve your simplified equations; write down this information clearly. If user defined functions are imported, or you developed your own code, write this detailed information down too. These details are important to the clarity of your writing, the validity of your results, and the credibility of your conclusions. Readers rely on the details to judge the quality of your work or to reproduce your results.

- **How the models are validated.**

"How much can I trust the model results?" Your readers may ask this question in their mind when they read your document. In the scientific and engineering world, seeing is believing. Data do matter.

Model validation is important and challenging to some numerical works. Nonetheless, you have to validate your models first before using them as a tool for your studies. If not by your own experimental data, you can validate your results with other models or the data in literature. Otherwise, it is viewed as over-confidence if you claim without evidence that your results are trustworthy.

In addition, you need to explain the error sources and their relative importance to the accuracy of your results. They may include the simplifications, assumptions, and so forth. You ought to point them out instead of leaving them to the readers to guess.

4.18.2 Experimental approaches

Repeatability is key to the description of experimental approaches. experimental approaches should be described in such way that the readers can repeat the work, if they decide to do so. The following general principles may assist you in achieving repeatability.

- Explain the experimental setup, if any, from overall to its components.
- Justify the designs of the setup and the components.
- Show the key specifications of the instruments and justify the choice of the instruments.
- Provide method for data analysis and, if feasible, expected results.
- Explain the errors and the uncertainties of the work.
- Report the preliminary trials, if any, to minimize the errors.

Example 4-52 shows the *Methodology* written in the sample paper (Fuller *et al.* 2012). It shows the level of details needed in describing their test procedures and data analyses. The writing contains precise information on the three *W's*: *where*, *when*, and *how* the experiments were conducted. This information allows readers to analyze the factors that may contribute to the uncertainty of results.

Example 4-52. Well-written methodology

Sampling Site Location and Sample Collection. RDI aerosol samples were collected at a site on the southern border of a dairy farm with about 1200 milking cows in the San Joaquin valley in California, one of the most intensively farmed regions in the United States. The sampling site is discussed in more detail in the Supporting Information and in Zhao *et al.* (2012)

An RDI, based on a design by Lundgren was used to collect size-segregated aerosol samples for approximately 2 weeks from 11 a.m. 5/25/2011 to 11 a.m. 6/7/2011. The fourth-generation RDI used in this study is described in detail in Zhao *et al.* (2012), 20 and only a short description is given here. Rather than collecting all particles on the same impaction spot during the sampling time, a Mylar strip on a slowly rotating drum is used as the impaction surface resulting in a time-resolved deposition of the impacted particles. Ambient air is sampled at a flow rate of 16.7 lpm, and aerosol particles are collected via impaction on greased (Apiezon L high-vacuum grease) Mylar strips. The Mylar is greased to increase the collection efficiency by reducing particle bouncing. The time resolution_is determined by the rotation speed of the RDI drums combined with the aerodynamic spread of sample deposition. This varies with the particle size cut, with smaller particle sizes giving higher resolutions. The RDI sampler consists of eight stages_collecting samples in eight size ranges (*i.e.*, stage 1, 10−5 μm; stage 2, 5−2.5 μm; stage 3, 2.5−1.15 μm; stage 4, 1.15−0.75 μm; stage 5, 0.75−0.56 μm; stage 6, 0.56−0.34 μm; stage 7, 0.34−0.26 μm; stage 8, 0.26−0.09 μm). Across all eight stages the type of deposition could be divided visually into two distinct sample types, first on the stages collecting particles with larger diameters (stages 1−5), brown colored particles follow a similar deposition variability with time. On stages with smaller diameter particles (stages 6−8) particle deposits were much darker and also showed a time variation similar to each other that was distinct from that observed on stages 1−5. Stage 3 and stage 8, corresponding to the size ranges of 2.5−1.15 and 0.26−0.09 um, respectively, were chosen as representative of the two deposition types and were investigated further in this study.

Sample Analysis. The organic composition of particles deposited on stages 3 and 8 of the RDI was analyzed with mass spectrometry using a new direct extraction technique, LESA. LESA is a mode of operation of the TriVersa NanoMate chip based nano electrospray ionization (ESI) source (Advion, Ithaca, USA). Samples are mounted onto a movable sample stage, and a small amount of solvent (1−10 μL) is then dispensed from a pipet tip on the surface of the sample (Figure 1). The micro liquid junction is maintained to extract analytes present at (or close to) the surface of the sample and to dissolve them into the small solvent extraction volume. The droplet is aspirated and subsequently sprayed via an infusion method utilizing the chip-based ESI22... All mass spectra shown here were recorded in positive ionization mode.

A section of each Mylar RDI sample strip was cut to size (20 mm × 70 mm) and mounted on a standard glass microscope slide before being analyzed by LESA at 20 points along a linear path at 1 mm intervals, which corresponds to the approximate liquid junction droplet size as shown in Figure 2. This corresponds to a sampling start time of about 19:30 on May 26, 2011 and an end time of 21:50,

May 28, 2011, with a sampling resolution of approximately 2 h 40 min. Blank extractions were taken from an area at the edge of the strip where no sample was deposited. Extraction volumes of 1 μL of solvent (80:20 methanol/water) were deposited at a height of 0.8 mm from the sample surface to form the liquid junction which was maintained for 30 s. This time allowed for analytes present at or close to the surface of the RDI strips to dissolve. Only very limited loss of solvent and resolution was observed due to droplet spreading. Many previous studies using LESA have repeatedly deposited and aspirated solvent onto the sample on a single extraction spot to aid mixing of the extracted sample into the solvent within a short contact time of typically 1-5s. However, this leads to sample loss through each deposition/aspiration cycle as a small amount of solvent is lost to the surface each time the sample is aspirated. However, a single deposition and aspiration reduces sample loss while increasing the contact time with the surface allows a longer period for analytes to dissolve and sufficient time for sample mixing through diffusion due to the small extraction volume. Contact times of over 30 s are less effective due to breakdown of the liquid junction.

An ultrahigh-resolution mass spectrometer (LTQ Velos Orbitrap, Thermo Scientific, Bremen, Germany) with a resolution of 100 000 at m/z 400 and a typical mass accuracy of ±2 ppm was used to analyze the organic compounds present in the samples following extraction by LESA. The resolution and mass accuracy of UHR-MS allows the identification of the elemental composition of unknown organic compounds. Samples were sprayed at a gas (N2) pressure of 0.30 psi at 1.8 kV in positive mode using a NanoMate nano-ESI source.

UHR-MS Data Analysis. For each extraction point on the RDI stripes mass spectra were recorded for a 1 min infusion duration in a mass range of m/z 50−500. Almost no peaks above m/z 350 were detected, and therefore only peaks below m/z 350 are discussed in the following. This mass range is a frequently observed characteristic of ambient aerosol composition. 25 The instrument was calibrated to within ±2 ppm using a standard calibration solution as prescribed by the manufacturer. Molecular formula were assigned within a ±2 ppm error and within the following restrictions: number of carbon-12 atoms = 1−20; carbon-13 = 0−1; sulfur= 0−1; sodium = 0−1; and the following elemental ratios of H/C = 0.2−3, O/C = 0−3, N/C = 0−1. Peaks and assignments not following the nitrogen rule or containing

C-13 were not considered further. In addition, peaks where no formula could be assigned within the restrictions mentioned above were also removed. Due to the low mass range of the detected peaks (below m/z 350) and the high accuracy of the instrument, multiple assignments are rare after considering the restrictions listed above. When several formulas satisfied all restrictions within 2 ppm, then the formula with the lowest mass error was assumed to be correct. Unfortunately, due to the generally low peak intensities of the identified species, MS/MS analysis for further structural identification was not possible. Only about 10−15% of the peaks contain a sulfur atom and are not further discussed here. See the Supporting Information for 46 more details on the data analysis procedure.

Comments

Overall, this example shows a well-structured *Methodology* section of an experimental work. The authors provided sufficient details and reasonable justifications for the methods used for data collection and data analysis. They recorded the procedure, in the sequence of time, as detailed as possible.

Some details might look trivial and unnecessary to you, but a well-trained experimentalist understands that something trivial might cause big errors. It is better to include the details, allowing your readers to evaluate the quality of your work.

The last (underlined) sentence shows the usage of *Supplemental Information*, which is briefly justified by the preceding sentence.

On the other hand, these paragraphs still have few minor errors in writing, even though the authors are native English speakers. First, they used emotional words (*Unfortunately*) and redundant conjunction (*and therefore*). In addition, using lists may enhance the smoothness of reading for the descriptions of the eight-stage *RDI sampler* and the *UHR-MS Data Analysis*. With the paper limit set by the publisher, the authors chose sufficient information (clarity) over perfect writing format, I believe.

4.18.3 Typical errors in writing *Methodology*

Lacking details is the most common mistake that non-native English writers make when they write their *Methodology* sections. Details are important for readers to evaluate the accuracy of results, validity of reasoning, and feasibility of data reconstruction. However, many published works lack details. This writing error may be attributed to omission of the following key information. (*See* Example 4-53.)

- Experimental setup and its components
- Test conditions and justification of parameters
- Models, suppliers, and specifications of devices and instruments
- Replicates of data point (enabling uncertainty/error analyses)
- Data analysis method (such as statistics software)
- Expected results (before presenting *Results*)
- Acceptance criteria (*e.g.*, *accurate*, *stable*, *equilibrium*)

Example 4-53. Poorly written methodology lacking details

2.2 Experimental method

Air was pumped and divided into two routes. One route of air was pumped through a water bath tower containing benzene solution to take gaseous benzene into a mixing chamber, and then the completely mixed gas was introduced into reaction chamber. The concentration of benzene in reaction chamber could be adjusted by a mass flow controller. Another route of air directly passed through a water bath container without benzene solution for carrying water vapor. The RH in the reaction chamber could be controlled from 20% to 60% by another mass flow controller. The air in the reaction chamber was circulated by a flow fan facing discharge gap to keep uniform relative humidity and stable discharge condition. The gas flow rate in the discharge channel was kept at 2.4 m/s, and the temperature of gas was 20±1 °C. After adding the catalyst to the system, the reaction was started by emerging the DC power supply. When the concentration of benzene in the reaction chamber reached a steady state, the benzene over the catalyst and organic glass chamber surfaces reached the adsorption-desorption equilibrium. *[Source: Ge et al. 2015]*

Comments:

In addition to language errors, the writing lacks details in the experimental procedure. The descriptions are primarily qualitative with little useful information. The writers omitted the details of the (underlined) materials, devices, and instruments. The justification of chosen parameters is also missing. Readers may ask the following questions. Is the *air* used pure? What kind of *water bath*, *chamber*, and *flow meter* are used? Why *20-60%* of RH and what would happen if RH is 80% or 10%? Why *2.4 m/s* instead of 2.5 m/s? What are the acceptance criteria for *equilibrium* and *complete mixing*? All these questions should have been answered in writing to enable data reconstruction by repeating the procedure.

4.19 Results and Discussion

4.19.1 Presenting results and discussion

Keep the *Objectives* in mind when you present the results and discussion. Present the results that are relevant to *the* objectives presented earlier. If you happen to obtain valuable results, but they are unrelated to your current objectives, you should save them for another publication or another division of the longer document (for example, a new chapter of the book). Mixing irrelevant ideas into one may cause confusion.

Always attempt in-depth discussion for publications that are aimed at new knowledge. Clearly state your opinion followed by logical arguments; do not assume that you readers will get it. The discussion begins with the most important and the most relevant to the objectives, followed by less important ones. You can compare your own results with those in literature and explain whether they are the same or different for discussion. If your findings are different from others, they deserve even further analyses. First, double-check your methodology to exclude any possible mistakes in data collection and analysis. Then, you can claim that your findings are new. Finally, support your claim with theoretical analyses, external data, and your own results.

All research has limitations. Presenting the weakness of your results, especially their uncertainties and errors, does not devalue your contribution to the field. By doing so, you actually gain respect from your peers and readers who highly value professionalism and clarity in writing. It also demonstrates that you clearly understand the subject from both sides and present them objectively.

Actively "engage" your readers when you write discussion. Writing is for the communication between you and your readers. Communicate with your readers as if they were sitting right in front of you. Anticipate the possible challenging questions that your readers may ask when they read your opinions. Then write down the answers to those possible questions, which can improve the clarity of your writing.

Example 4-54 shows a well-written *Results and Discussion* section (Fuller *et al.* 2012), which follows the order of *results, explanation, validation,* and *in-depth analyses.* A few journals require separating results from discussion, but it should not affect the pattern of presentation.

Example 4-54. Well-written results

……Figure 4 shows the O/C and N/C ratios of the most intense peaks in the mass spectra as a function of their molecular mass from both stages. For clarity only the most intensive 200 peaks in the mass spectra are shown here, which correspond to about 80% of the total ion intensity. The molecular formulas from all 20 extraction points were combined and their intensities from each point summed. …… Stage 3 is shown in blue and stage 8 in red, superimposed on top of the stage 3 data. The elemental ratios of O/C and N/C ratio were found to vary with particle size fraction (between the stages) and with molecular mass range.

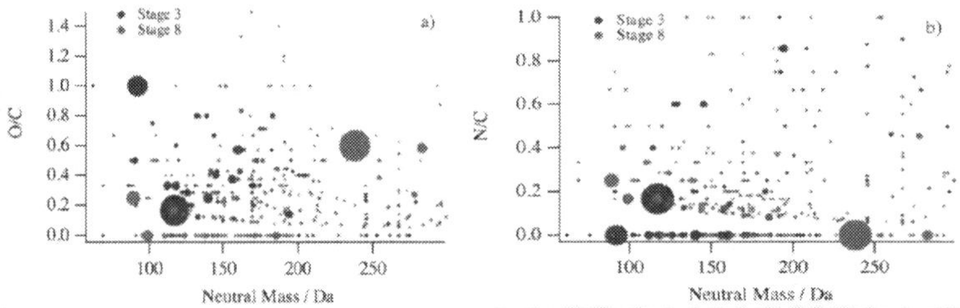

Figure 4. Elemental ratios of formulas for the most intense 200 peaks in the mass spectra in both stages (3 and 8) as a function of their neutral mass: (a) O/C ratio; (b) N/C ratio. Symbol size reflects peak intensity.

Comments:

The writer effectively integrates the visuals into the body text. The text introduces the elements (legends, contents, *etc.*) in the figures, which are visualized. Readers can capture the main ideas by looking at the visuals or the text only. Example 4-55 shows presentation of results with *explanation* and *validation* preceding *in-depth analyses.*

Example 4-55. Explanation and validation of results

......Figure 4a shows 15% more compounds containing only C, H, and O among the 200 most abundant species in stage 3 than stage 8 (165 compounds compared to 141). For these CHO species in stage 3, 69% of the top 200 peaks by number and 87% by intensity have a neutral mass of <200 Da. In comparison, for stage 8 only 57% by number and 45% by intensity are in this mass range. This is due to a larger number of compounds in stage 8 with no oxygen content, but only C, H, and N, indicating the presence and importance of amine type species in this submicrometer particle size range. The similarity of the nitrogen content between the stages is due to this increased abundance of CHN compounds in stage 8 being balanced by presence of more oxidized nitrogen species (CHNO) in stage 3.

Figure 4b shows that [text omission] ......; however, for stage 3 for a neutral mass of <200 Da a large number of high-intensity compounds are present on the baseline with an N/C ratio of zero, indicating that *[text omission]* ......

For all compounds below m/z 150 (181 compounds) in the large size fraction (stage 3) the N/C ratio is clearly correlated with the overall particle loading and wind speed (Supporting Information Figure S2). This strongly suggests that compounds in the low-mass region with high nitrogen content may be soil derived. Soil organic compounds are known to have a large number of nitrogen-containing compounds. The sampling location close to a dairy farm could also partly explain the high N/C ratio of the large particle fraction under high-wind conditions, due to the high N content of animal waste. This correlation is not as strong if all compounds up to m/z 350 are considered.

Comments:

The contents introduced by the underlined words and phrases (*due to, explain, suggests, etc.*) explain the results. The explanations help readers *validate* the results before deducing new knowledge by *in-depth analysis* that follows.

You cannot stop at *explanation*; *in-depth analyses* must be provided to meet your research objectives. Most beginner writers simply stop here and move on to other results or jump into conclusions. Skipping *explanation*, *validation*, or *in-depth analyses* is a typical mistake in poor technical writing. It undermines the logic in argument and devalues the writer's work. (*See* Example 4-56)

Example 4-56. In-depth analyses of results and limitations

As shown above it is often difficult to identify groups of compounds originating from a common source, and correlations of peak intensities over time allow the possible identification of compounds with a common or similar atmospheric history. Thus, for each identified formula the variation of intensity across the 20 sampling points was compared to every other formula in the entire data set. Linear regression analysis was used to indicate correlations, and any correlation with an R^2 value of greater than 0.8 was defined as statistically significant. This analysis assumes that matrix ionization effects are constant throughout the RDI stripe due to the hydrocarbon grease applied to the Mylar strips. This resulted in a number of sets of correlating species for stage 3. Three of these correlating sets of species are shown in Supporting Information Table S1, which gives for each set the formula and the total peak intensities across all 20 extraction points. Figure 7 shows the variation in the averaged intensity for all the species in each set for each extraction point, which was subsequently normalized to the maximum average intensity extraction points. Series 1 contains both CHNO and CHN species, and series 2 contains only CHO compounds. Both series show a negative correlation to the brown-colored regions in stage 3 (Figure 2a). In contrast, series 3 contains CHNO compounds only and shows a positive correlation with the brown deposit (extraction points 9/10 and 17, Figure 2a). Series 3 also correlates with the overall N/C ratio and wind speed as shown in Supporting Information Figure S2, again suggesting that these compounds may be soil-derived and from local (farm) sources. No series of correlating species were identified by this method in stage 8 indicating a more complex composition and atmospheric history of the small-particle fraction.

Comments:

This paragraph begins with the method used for in-depth analyses of the validated results, followed by new knowledge of the organic compounds, and ends with limitations of the work *(due to complexity)*.

Example 4-54 through Example 4-56 illustrate the sequence of results, explanation, validation, and in-depth analysis. keeping this presentation pattern in mind ensures the logics of your writing.

Note:

Stating explicitly an opinion is considered inappropriate or unnecessary in certain cultures, but readers expect the writer's point of view to be stated clearly for engineering publications in English. However, the statement should be followed by logical arguments.

4.19.2 Writing errors in results and discussion

Writing errors may appear in the *Results and Discussion* section when you neglect the *result-explanation-validation-analysis* pattern introduced in Section 4.19.1. There are various errors, depending on the writer's training and experience in writing, and the following often appear in engineering publications.

- Integrate poorly visual into text (Example 4-57; Example 4-58)
- Present results without in-depth analysis (Example 4-59)
- Interpret visuals with errors (Example 4-60)
- Forget error bars in results, if applicable (Example 4-60)
- Lack convincing argument in discussion
- Refuse to accept limitations

Example 4-57. Poor integration of visuals into text (1)

[Beginning of the paragraph] Figure 6 shows the sketch of the room and the positions for eight measuring poles (*citation*). Heat sources, including human simulators, lights, computers, are present in the real office as well as in the simulations. Contaminant sources are the human simulators. Table 1 summarizes the parameters of the supply and return air including inlet temperature, outlet temperature, inlet velocity and outlet velocity. *[End of the paragraph]*

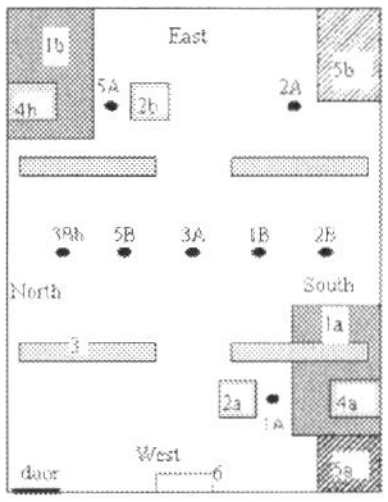

1—table (a, b), 2—human simulators (a, b), 3—lights, 4—computers (a, b), 5—cabinets (a, b), 6—displacement diffuser

	Displacement case	Grille case
Inlet temperature (°C)	15.88	18.49
Outlet temperature (°C)	24.80	24.16
Inlet velocity (m/s)	0.3	1.55
Outlet velocity (m/s)	0.29	0.47
Ventilation rate (m³/s)	0.0562	0.0674

Table 1 [*Table caption*]

Example 4-58. Poor integration of visual into text (2)

Wrong: Fig. 5 shows that the model agrees well with the experiments.
Right: Figure 5 shows that the maximum difference between the model and experimental data is 0.5%.

Example 4-59. Results without validation or in-depth analysis

Fig. 8 shows the percent changes in the aerodynamic force coefficients and rolling moment coefficients due to rain impact for various rain intensities with certain yaw angle. Different colored bars shown in the figure represent the percent changes in coefficients under different rain intensities. From Fig. 8(a)~(b), it is clearly that both the percent changes in coefficients of aerodynamic drag force and side force are in the character of ascent as the amount of rainfall is up for all the yaw angle calculated in this work. Fig. 10(a) indicates that the drag coefficients increases fall in the range from 1.5% to 38.6% across the yaw angle. And the relative high drag coefficients increases occur at large yaw angle: when the yaw angle equals to 30.97°or 35.76°, the drag coefficients increases are within the range of 13.0% - 39.0%. Fig. 8(b) indicates that the percent changes inside coefficients vary from 0.5% to 9.2% across the yaw angle. Similar to those of drag coefficients percent changes, the relative high drag coefficients increases occur at large yaw angle: when the yaw angle equals to 35.76°, the side force coefficients increases have a value between 2.5% to 10.0%. Fig. 8(c) shows that there is not a generalized regularity between the rain intensity and the lift force coefficients across the yaw angle range. However, distinct regularity still can be seen when the yaw is very low or relative high: for the low yaw angle which equals to 6.85°, the percent changes in lift force coefficient has an upward trend with the increasing of rain intensity; on the contrary, for the high yaw angle which equals to 35.75°, the percent changes in lift force coefficient fall down as the rain intensity increasing. For the rolling moment coefficients, they increase as the rain intensity gets heavier across the yaw angle range as shown in Fig. 8(d). And the rolling moment coefficients increases from 1.8% - 10.0% are derived for the various rain fall rates. Furtherly, the relative high rolling moment coefficients increases occur at low yaw angle: when the yaw angle equals to 6.85°, the rolling coefficients reaches 4.0% - 10.0%.

Fig. 9 shows…… *[Last paragraph is written with the same style.]*

4. Conclusions

Comments:

This is from a rejected manuscript that I reviewed. The entire section roughly translates the contents from the visuals to the text. The writer drew the *Conclusions* without data validation or in-depth analysis. As a result, they are weak conclusions that undervalue the original work.

Example 4-60. Wrong interpretation of visual

[Beginning of the paragraph] From Fig. 3, it can be seen that the benzene removal efficiency increased with increasing discharge power both in NTP or CPMC. The influence of discharge power on benzene removal efficiency can be ascribed to the influences of voltage and current……

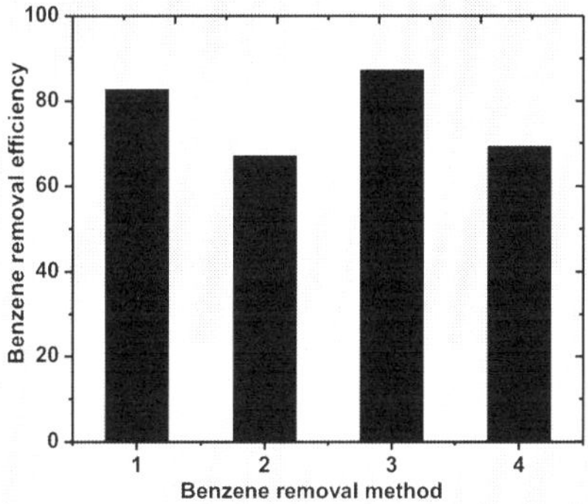

Fig. 3 The effect of discharge power on benzene by plasma only and plasma-MnO_2 systems. 1.NTP, 9 W; 2. NTP, 4.5 W; 3. CPMC, 9 W; 4. CPMC, 4.5 W

Comments:

This is part of a manuscript that I reviewed for a journal. The writing has multiple mistakes in both language and technical contents. First, there is a mismatch between the figure and the body text. Fig. 3 does not support the underlined text, which is qualitative. Second, the data are questionable when they are presented without error bars. In addition, the visual lacks clarity because of the poor format (see 9.10). Clear visuals allow the readers to get most of the message in the visuals without heavy reliance on the text, and *vice versa.*

Admittedly, this is an extreme case. What the writer submitted is a *crude draft.* You should never submit a manuscript like this to any journal for external review. Be mindful that reviewers are professional volunteers. The reality is that manuscripts like this one are submitted to journals for review, and that some managed to get their papers accepted for publication (*See* Example 4-35). After reading this book, you should avoid this type of blunder.

4.20 Drafting Conclusions

Conclusions of a document need to answer the ultimate question: have you achieved the *Objectives*? The conclusions should be based on the results, interpretations, and in-depth analyses presented before this point. The conclusions point all the evidence at one spot, which may include the key findings, new knowledge, and recommendations. The conclusions must be drawn from the evidence that is presented in the same document, including work reported by others.

A well-written *Conclusions* section is closely linked to the *Objectives* presented earlier in the same document. State with clarity and conciseness how the objectives are achieved. *Recommendations*, which are optional, can be presented at the end of the *Conclusions* section. Example 4-61 shows the conclusions drawn from the results and in-depth analyses (Fuller *et al.* 2012).

Example 4-61. Well-written *Conclusions*

[First paragraph of the Conclusions] We have developed a new technique to analyze the organic composition of atmospheric aerosol on a molecular level with a high time resolution by combining a new extraction and ionization technique, LESA, with RDI samples and ultrahigh resolution mass spectrometry. LESA facilitates analysis without time-consuming sample preparation or extraction and also allows direct extraction from a sample surface with high spatial resolution, which when combined with RDI samples results in highly time-resolved information of ambient aerosol organic content.

[Second paragraph of the Conclusions] Ultrahigh-resolution mass spectrometry was used to determine the chemical composition of organic compounds and to investigate their changes in intensity over the sampling period in two aerosol size ranges. Changes in the chemical composition could be related to changes in meteorological conditions; for example, the signal intensity of a subset set of nitrogen-containing compounds was found to be associated with the local wind speed suggesting that these compounds are largely of local origin. Groups of compounds were identified that showed similar trends in abundance over the sampling period indicating that these compounds originate from the same source. This was demonstrated by the

identification of homologous series, which are likely associated with biomass burning sources. A further characterization of the structures of the compounds, e.g., by performing MS/MS experiments, was not possible due to the generally low peak intensities. This could be overcome in future studies by a lower rotation speed of the RDI during sample collection, which would allow for higher sample loadings and might allow sufficient peak intensities to perform MS/MS experiments.

[Last paragraph of the Conclusions] While mass spectrometry is widely used to identify organic aerosol content, the combination with LESA and RDI sampling allows a much higher temporal sampling resolution than would be possible with other off-line techniques allowing one to identify marker compounds and to follow atmospheric processes in detail.

Comments:

The first paragraph states that the researchers met the first objective, *developing a new technique*, stated earlier (Example 4-49).

The second paragraph concludes the second objective, *characterizing the aerosol particles*, and so forth.

The last paragraph in the Conclusions seems to be isolated from other paragraphs. The authors might hope to address a reviewer's comments or concerns, or they wanted to emphasize the practical values of the new techniques. In terms of writing, however, this single-sentence paragraph should be merged into another one that has the same controlling idea. (*See also* 5.1.2 Paragraph length.)

4.21 Acknowledgements

You can acknowledge your financial sponsor(s), and people who contributed to your publication but not significant enough to become the co-author(s). Example 4-62 shows the *Acknowledgements* section in a journal article written by Gioia et al. (2011).

Example 4-62. Acknowledgements

Michael C. Dameron, John C. Griffith, Je I. Ryu, William F. Burgoyne, Federico Scholcoff, and especially Paulo Zandonade provided much help with the experiments. We thank James W. Phillips (University of Illinois, Urbana, IL) for the photograph of Fig. 1A. We thank

Zhongchao Tan (University of Waterloo, ON, Canada) for the photograph of Fig. 1D. This work was financed in part by National Science Foundation Division of Materials Research Grant 10-44901. P.C. acknowledges support from The Roscoe G. Jackson II Research Fellowship.

4.22 References and Bibliography

A list of references appears near the end of the document. Listing references in a consistent format allows readers to locate and access the works cited for further information on the subject. (*See also* 9.12.) It also allows the readers to validate the data used by the authors to support the conclusions in the original document. Equally important, it helps avoid plagiarism by giving proper credit to the original creators of the knowledge.

A bibliography is a list of all the sources that are used to support the writing of the document. The bibliography includes additional resources for further reading by some readers who may need additional information. It usually includes the references and those not cited. The sources include books, articles, theses, dissertations, reports, creditable online sources, and other works. They are normally listed alphabetically by the authors' last names. *See* 9.12 for more information.

4.23 Back Matter

4.23.1 Appendices and Supplemental Materials

Supplemental materials, which are non-essential to the primary readers, are located at the end of the documents. They are labeled as *Appendices* in long documents and *Supplemental Information* (or the like) in short documents. Many prestigious journals, like *Nature* and *Science*, require authors to use supplemental information because of page limit. The supplemental materials may include detailed derivation of equations, non-essential visuals like maps, coding for statistical analyses, and credentials of the main researchers. Appendices and Supplemental Materials are optional components for secondary readers.

Example 4-63 shows the content and the format of the *Supporting Information* in the final printed article written by Fuller *et al.* (2012).

Example 4-63. Supporting Information in text

Supporting Information

Additional information as noted in text. This material is available free of charge via the Internet at http://pubs.acs.org.

4.23.2 Glossary

Glossaries are lists of definitions of specialized terms used in the documents. Glossaries are optional to short documents, but they may appear after bibliographies or appendices in long formal documents. Alphabetically arrange the entries of a glossary for clarity. A glossary seems to have the same function as an index, but the glossary has much more information than the index.

4.23.3 Index

Index, if any, is located at the very end of the long document. It is a list of the keywords and phrases used in the document. Like bullet points, the index can also include sub-entries. Unlike glossaries, however, indexes do not include definitions. Indexes are available in most books, but they are not needed in articles, short reports, or theses.

Begin compiling the index when the final manuscript is ready. At this point, you can read through your document from the cover page to the end. Note the key terms and their page numbers each time they appear. The entries in the index can include keywords and phrases in the table and figure titles. Arrange the index entries alphabetically with page numbers so your readers can locate them easily.

It is a time-consuming task to manually compile an index. Nowadays, software is available for this kind of tasks. For example, the American Society of Indexers (www.asindexing.org) lists software available to indexers for this process. Most word processing software, such as *Microsoft Word*, has this built-in function.

4.24 Getting Ready for Revisions

Now you have finished the first draft of your document. Your manuscript may be rough and have many writing errors, and this is normal to most writers, regardless of their experience. That is why the next step, revision, is critical to successful writing too.

Begin revision as soon as you finish your first draft. Revision is time consuming and you may repeat it three or more times for the entire document. Your efforts can gradually refine your manuscript to the level of your satisfaction.

Concentrate your mind on three *C's: clarity, coherence, and conciseness* when you revise the document. Read through the document and ensure smooth transitions between paragraphs without interrupting your flow of thought. Pay attention to the logics of your writing, and make sure that your evidence and argument really support your statements and conclusions. Your writing should deliver exactly what you expected to share with your readers.

Meanwhile, consider grammar, spelling, structure, punctuation, and so forth when you edit the draft manuscript. Grammar sets the rules you need to follow in writing (and speaking). Grammatical correctness is the basics of academic writing. Precise wording is also important to reducing ambiguity (*See* 7. Words and Phrases).

Carefully identify and remove the redundant materials from the draft to improve conciseness (Example 4-64). Unlike an insurance policy or a legal document, professional writing is characterized with clarity and conciseness, especially after three rounds of editing.

Example 4-64. Removal of redundant materials in editing

Redundant: The *application* of a heterogeneous photocatalyst has been successful in two applications. *[written by a native speaker]*

Correct: ~~The application of a h~~Heterogeneous photocatalyst has been successful in two applications.

Correct: The application of a heterogeneous photocatalyst has been successful in *two industries.*

Finally, Example 4-65 illustrates the importance and effectiveness of revision. This book begins with the text *after revisions.* (*see* pg. 1). These two paragraphs are derived from one single paragraph in the *first draft* of this book. The comparison shows how the revisions improve the coherence of writing by ensuring the singleness of idea in each paragraph. The revisions also follow the guidelines of paragraph structure and sentence structure (*See* Chapters 5 and 6). Meanwhile, the refinements in grammar and spelling improve the clarity and conciseness of my writing (*See* Chapters 7 and 8).

Example 4-65. Comparison between first draft and revised manuscript

First draft	After revisions
Written communication is a type of skill, albeit an important one. We write at work everyday because writing is a great extension of our voices, conveying our thoughts to countless people that we may never meet with in our life. It reveals our intelligence of thinking, ability of using words, level of education, and so forth. Good writers are normally creative people with brilliant ideas, a trait that can make capable writers excel in their career development. Regardless of your work, we need to write almost everyday.	Writing is a type of communication skills, albeit an important one. Written communication is more than fluent speaking or a good command of grammar, spelling, and punctuation. Writing is a complex task requiring training and practice of many techniques, such as organizing ideas logically, constructing sentences and paragraphs coherently, presenting with appropriate tones, formatting in a stylish manner, and executing in an ethical and professional matter. Written communication reveals our intelligence of thinking, ability of using words, level of education, and so forth. Good writers are normally creative people with brilliant ideas, a trait that can make capable writers excel in their career development. Regardless of your jobs, you might need to write daily. Writing is a great extension of your voices that conveys your thoughts to many people in the world. Some of them may never contact or meet you.

~~~

The guidelines in the following chapters can help you improve in paragraph structures, sentence constructions, word choices, visual integration, and punctuation marks. The document after multiple revisions should be free of ambiguity and confusion.

Finally, revision is neither proofreading nor formatting. Revisions and proofreading require different skills and techniques, even though they are occasionally used interchangeably by mistake. (*See* 10 for proofreading.) Refrain from formatting the contents when you revise the document. Very likely the formats change during your revision. Leaving formatting to the near end, before proofreading, will not only saves you time but also helps you focus your attention on the important.
~~~

5 Paragraphs

A paragraph is formed by sentences that are related to the single controlling idea of the paragraph. The sentences in the paragraph must be organized with coherence. A paragraph could describe a procedure, compare two or more ideas, classify a family of technology into categories, analyze the cause of effects, draw conclusions from analyses, make recommendations for actions, and so forth. It develops the coherent idea, provides a convincing argument, and signals a new topic. All paragraphs share certain common structural characteristics, and this chapter introduces the techniques for writing a well-organized paragraph.

5.1 Paragraph Structures

As illustrated in Figure 4-1, a paragraph normally consists of three key elements: a topic sentence, body sentences, and a concluding sentence. The topic sentence is a statement that prescribes the controlling idea of the paragraph. Following the topic sentence is the body of evidence that supports the statement. It elaborates on the statement using logical arguments, examples, analyses, and other supporting evidence. The paragraph ends with a concluding sentence, which reminds the readers of the main point of the paragraph. Each of the sentences in the paragraph plays an important role in communication between you and your readers.

5.1.1 Topic sentences

The topic sentence declares the single controlling idea of the entire paragraph. The topic sentence unifies the controlling idea of the paragraph and guides the writing with the order of the sentences to follow.

A topic sentence is optional to a paragraph that continues developing the same idea as that in the previous paragraph.

It is important to put the topic sentence at the beginning, mostly the first sentence, of the paragraph because most readers look to the topic sentence to determine the single idea of that paragraph. Occasionally, a transitional sentence proceeds the topic sentence, which links the current paragraph to the previous one.

The *statement-evidence-conclusion* structure is typical to writing in English, but not necessary for works produced in another language. Therefore, it may take some practices for you master the paragraphing techniques.

Example 5-1 and Example 5-2 show imperfect but acceptable paragraph structures. The contents are from the same article written by Fuller *et al.* (2012) unless stated otherwise. In contract, Example 5-3 through Example 5-8 illustrate problematic paragraphs with various errors.

Example 5-1. Paragraph with statement and two supporting examples

There are a number of analytical−chemical techniques that allow measuring the composition of aerosol particles with high time resolution. Online aerosol mass spectrometry (AMS) techniques, for example, allow for very highly time-resolved particle composition studies. However, such measurements are demanding with respect to manpower and other resources and are usually performed only for a few weeks at a specific site. Thus, such measurements rarely provide insight into long-term trends of particle composition. In contrast, aerosol samples collected on filters or impactors over long time periods are more readily available but have mostly a rather low time resolution of the order of a day or more, and their chemical analysis usually involves time-consuming.

Comments:

[This is the same paragraph as in Example 4-46.] The paragraph begins with a topic sentence (*There are a number of analytical−chemical techniques*) followed by two examples as supporting evidences. One example is *(AMS) techniques* and the other, *aerosol samples collected on filters or impactors.* The closing sentence is omitted, and this type of short structure is acceptable in spite of its imperfection.

Example 5-2. Paragraph with statement followed by evidence

Figure 4 shows the O/C and N/C ratios of the most intense peaks in the mass spectra as a function of their molecular mass from both stages. For clarity only the most intensive 200 peaks in the mass spectra are shown here, which correspond to about 80% of the total ion intensity. The molecular formulas from all 20 extraction points were combined and their intensities from each point summed.

Example 5-3. Multiple ideas in one paragraph

[Begins Introduction] In the last few decades high speed train (with speeds up to x kph) have been designed and built in a number of countries. Coupled with this increase in speed is a general tendency for the weight of the train to reduce. Hence, the risk of overturning potentially increases, especially when a train travels during periods of bad weather such as strong cross winds, rainstorms or sandstorms, *etc.* Most of current studies were concerned with train under cross wind, including aerodynamic characteristics and running stability with various factors such as configurations of vehicles and infrastructures, the yaw angle and gust shape, *etc.* And some further address the primary cause of the aerodynamic forces such as the flow field. The major methods to study a train under cross wind contains full-scale tests, wind tunnel tests, and numerical simulation. Panel methods and Multi-body simulation method were also applied in train's stability analysis. However, research on high-speed train traveling under severe bad weather such as thunderstorms (wind-rain condition), sandstorm, etc. are few.

Comments:

This paragraph is part of a manuscript that a graduate student asked me to review. First, the paragraph needs a topic sentence at the beginning. The writer put the current scope (*high speed train*), scopes of earlier works (*most studies… and some further…*), and other research methods (*The major methods to study a train*) into one single paragraph. The rapid pace of writing causes idea clutter. The revisions, following the *statement-evidence-conclusion* paragraph structure, help improve coherence in writing.

Suggested revisions:

In the last few decades high speed trains (with speeds up to x kph) have been designed and built in a number of countries. For example, Germany built 100 from 1990 to 2012 with 10 more in the pipeline [citation]. There are even more in China with a total number of 200 between 2000 and 2010 [citation]...... High speed trains become part of the model mobility system.

Coupled with this increase in speed is a general tendency for the weight reduction of high speed trains ~~to reduce~~. The weight of today's typical vehicle is only 20% of that 20 years ago. In 1990, a typical vehicle weighted about 1000 tons and now it is only 200 tons [citations]. Weight reduction is important to train ~~speed~~ manufacturing and operation.

This reduced weight comes with ~~Hence,~~ the increased risk of overturning ~~potentially increases~~. *[add evidence to support the statement.]* It is especially true when a train travels ~~during periods of~~ in bad weather such as strong cross winds, rainstorms, or sandstorms~~, *etc*~~.

There have been a number of studies in high speed train under challenging conditions. Most of them ~~current studies~~ were concerned with trains under cross wind. Factors of concern include aerodynamic characteristics and running stability with various ~~factors such as~~ the configurations of vehicles and infrastructures, the yaw angle, and the gust shape, *etc.* ~~And some further~~ Others address the primary cause of the aerodynamic forces such as the flow field. The major methods to study a train under cross wind ~~contains~~ include full-scale tests, wind tunnel tests, and numerical simulation. Panel methods and multi-body simulation method were also applied ~~in~~ to train's stability analysis.

To the authors best knowledge, however, there is little research on high-speed train traveling under severe weather, such as thunderstorms, wind-rain condition, sandstorm~~, *etc*~~. *[add more evidence]*

Comments:

The suggested revisions are focused on dividing a long paragraph into multiple paragraphs. Each of the new paragraphs has one single controlling idea with the added text (underlined) or evidence to be added. The original writing also needs refinements in language.

Example 5-4. Statement at the end of a problematic paragraph (1)

2.1. Cities

All 16 cities included in CAPES are involved in our study, namely Anshan, Beijing, Fuzhou, Guangzhou, Hangzhou, Hong Kong, Lanzhou, Shanghai, Shenyang, Suzhou, Taiyuan, Tangshan, Tianjin, Urumqi, Wuhan and Xi'an. These cities are geographically widely spread throughout China and vary greatly in climate. Their locations are shown in the Figure 1. *[Source: A draft reviewed in 2013]*

Suggested revisions (*statement-evidence-conclusion structure*):

Figure 1 shows the locations of the cities in this study. All 16 cities included in CAPES are involved in ~~our~~this study, namely Anshan, Beijing, Fuzhou, Guangzhou, Hangzhou, Hong Kong, Lanzhou, Shanghai, Shenyang, Suzhou, Taiyuan, Tangshan, Tianjin, Urumqi, Wuhan and Xi'an. These cities are geographically widely spread throughout China and they vary greatly in climate.

Example 5-5. Statement at the end of a problematic paragraph (2)

The photocatalysis begins when solar energy meets the TiO_2 semiconductor, exciting an electron of a TiO2 molecule from the filled valence band to the empty conduction band. The energy required to excite the electron from the valence band to the conduction band is known as the band gap. When electrons migrate to the surface of the TiO_2 material and are trapped at the edge of the conduction band there, they serve as reduction centers, donating electrons to other acceptors. Similarly, the holes left in the valence band serve as oxidizing sites. Figure 1 provides a schematic diagram of the photocatalysis process.

Suggest revisions:

Relocate the last sentence to the beginning of the paragraph. It should bring the controlling idea up front. Note that this is a paragraph in a graduate-level course project report submitted by a native English speaker. It further emphasizes the difference between written and oral commination skills.

Example 5-6. Statement at the end of a problematic paragraph (3)

Three long-span bridge deck cross sections are taken as examples in the present paper. They are named as sections G1–G3. Their geometrical features are shown in Figure 1, in which *B* is the chord length. Every model has two degrees of freedom, namely vertical translation *h* and rotation about its center. Section G1 can be treated as a streamlined structure, but G2 and G3 are typical bluff bodies with sharp edges. Particularly, the section G3, which has infamous aerodynamic instability because it is the prototype of the Tacoma Narrows Bridge in the USA, was destroyed in 1954 by a steady wind with the small velocity of 20 m/s. So, it is necessary to investigate these structures. *[Source: Bai et al. 2013]*

Example 5-7. Find the mistakes in this paragraph

[This is the second last paragraph of Introduction.] It is well known that DC streamer discharge has many advantages such as high energy efficiency, discharge stability as well as high efficiency in the generation of chemical active species. More importantly, O_3 and NO_2 production in this discharge mode are significantly less than that of in the first two discharge modes. Hence, the combination of streamer discharge and MnO_2 catalyst should be a promising method for indoor air VOCs removal. Until now, only a few studies concerned on the feasibility of applying streamer discharge combined with MnO_2 for the removal of indoor air VOCs. *[Source: A manuscript I reviewed and rejected]*

5.1.2 Paragraph length

A paragraph should provide the readers with manageable subdivisions of thought. Otherwise, the writing appears to be disorganized. A series of short, undeveloped paragraphs break a single idea into several pieces, whereas a paragraph that is too long loses its conciseness and coherence (*See* Section 0 above). For this reason, it is generally not acceptable to write a single-sentence paragraph.

Note:

A one-sentence paragraph is rare, but it is acceptable for

a) a transitional sentence between two paragraphs.

b) a conclusion in a list.

Example 5-8. Single-sentence paragraph

While mass spectrometry is widely used to identify organic aerosol content, the combination with LESA and RDI sampling allows a much higher temporal sampling resolution than would be possible with other off-line techniques allowing one to identify marker compounds and to follow atmospheric processes in detail.

Comments:

As explained in Example 4-61, this paragraph appears at the end of the body of the article. Therefore, it is not a transitional paragraph. You should avoid this type of single-sentence paragraph in your writing.

5.2 Lists in Paragraphs

Appropriate use of *list* is often more effective than a long paragraph for the presentation of certain information, for example, materials used, parts needed, steps in experiments, and key points for conclusions and recommendations.

The list entries may be words, phrases, clauses, and simple sentences. They can be listed vertically using numbers, letters, bullet points, or their combinations. Numbered lists can save readers time by focusing on steps of sequence. Use bullets for items without rank or sequence.

The following techniques are useful to the creation of effective lists.

- Maintain parallel structure for the entire list.
- List only comparable items of balanced significance.
- Capitalize the first word in each entry.
- Avoid commas or semicolons at the end of each item.
- Use ending punctuation (*e.g.* periods) at the end of each item if the list consists of complete sentences.
- Do not use "*and*" at the end of the second last entry.

Nonetheless, avoid overuse of *lists*. An unnecessarily long list with too many items loses its effectiveness. In addition, make sure that you provide the context of the list. It would be difficult for your readers to follow your ideas if the paragraphs consist entirely of lists.

5.3 Quotations in Paragraphs

You can *occasionally* use a direct or indirect quotation to stimulate interest in your subject (*See* the beginning of Chapter 7). You may use quotations for drafting but use as few quotations as possible in the final version of your document. It is better to rephrase the direct quotations in your own language and give credit to the original author(s).

Paraphrase is commonly used for indirect quotations, which is usually introduced by the word *that*. Indirect quotations appear to be your own writing, but not. You need to use the references and in-text citations to avoid plagiarism.

Example 5-9. Direct and indirect quotations

Direct: Stephen Hawking once said, "Quiet people have the loudest minds."
Indirect: Stephen Hawking once said *that* quiet people have the loudest minds.

5.4 Visuals between Paragraphs

It may be a *cliché* to say, "A picture is worth a thousand words," but visuals do have power. Effective communication does require a multisensory experience. Readers notice visuals first when the visuals are placed besides sentences and paragraphs. Colorful and bright visuals are more noticeable than ordinary ones. In addition, a nice illustration in the crowd of words gives the readers a visual relief.

Technically, visuals can convey messages more effectively than texts alone. For example, a schematic diagram can show how a complex system works, which may take multiple paragraphs or pages to explain. Charts and tables can also describe the relationship between numbers more clearly than words do. You are encouraged to use appropriate visuals to enhance your writing.

5.4.1 Placement of visuals

Visuals are often placed in the main body of text right after their first explanations. However, visuals should not precede their first mention.

There are several best practices in integration of visuals with the text. In text, you can refer to the visuals by their figure or table numbers. Capitalize the word *figure* or *table* because (*i.e.,* Figure, Table) because it refers to a specific section of the document. Depending on the format guideline, you can spell out the word of figure or use their abbreviation. Same rules apply to the cross-reference of equations, although equations are not normally considered as visuals. However, no abbreviation is used for table. Table 5-1 summarizes the spelling of cross-reference of figures, tables, and equations in academic engineering writing.

Table 5-1. Cross-references of visuals and equations

Type	Cross-ref.	Meaning
figure	Figure 6	The sixth figure
	Fig. 6	Same as Figure 6
	Figure 6-8	The eighth figure in Chapter 6
	Fig. 6-8	Same as Figure 6-8
	Figures 6-8	Figures 6, 7 and 8
	Figs. 6-8	Same as Figures 6-8
table	Table 6	The sixth table
	Table 6-8	The eighth table in Chapter 6
	Tables 6-8	Tables 6, 7, and 8
equation	Equation 6	The sixth equation
	Eq. 6	Same as Equation 6
	Equation 6-8	The eighth equation in Chapter 6
	Eq. 6-8	Same as Equation 6-8
	Equations 6-8	Equations 6, 7, and 8
	Eqs. 6-8	Same as Equations 6-8

5.4.2 Creating visuals with clarity and conciseness

Like with the text, the visuals also should be created and presented with clarify, conciseness, and consistency. Visuals and the related text most often support each other. The visuals aid readers to understand the text, but they cannot replace the text. Meanwhile, the contents of the visuals appear in the text that explains the visuals. You can follow these general guidelines on visuals when you edit your draft. (*See also* 9.10)

Note

Do not use copyrighted materials without written permissions from the copyright holders. Reproduced works also require permissions. This ethics rule applies to both printed and online publications.

5.4.2.1 Contents in visuals

When you create the visuals, include only information that is needed to support your arguments. Eliminate unnecessary contents in visuals. You can place lengthy, detailed visuals into appendices or supplemental materials, if they are non-essential to the body of the text. That ensures the clarity and conciseness of the main document.

You can use the following techniques to avoid visual clutter.

- Use text with compatible typeface and font.
- Use symbols that are known globally (*e.g.* % for percentage, °C for degree Celsius); otherwise, spell out the words if the symbols are defined for your document only.
- Specify the scales and the units of measurement (*e.g.* axis's for line graphs).
- Use terminologies that are consistent with the main body.

5.4.2.2 Captions of visuals

The function of captions to the visuals is like the titles to the main documents: captions are titles that describe their visuals. The captions should capture key messages contained in the visuals. The figure captions often appear below the figures, and the table captions above the tables.

You also need to number the figures and the tables. They follow two different orders of sequence. Their numbers and captions will be referred to in the List of Tables and the List of Figures, if applicable. (*See also* Table 5-1 for cross-reference of visuals.)

5.4.2.3 Labels in visuals

Labels are necessary to the clarity and the conciseness of visuals. In schematics, for instance, you need to label the parts or components that you explain in the text. In a map, as another example, you need to label the key streets and the locations that are relevant to your research. In-visual labels are generally preferred. You may put labels inside simple visuals with clear and consistent texts. (*See* Figure 5-6 for example.) On the other hand, make sure the labels do not clutter the visuals. For instance, you may label parts of a complex schematic diagram with numbers and list their texts at the bottom, below the caption, or on the side of the schematic diagram.

5.4.2.4 Horizontal direction of text in visuals

Position the explanatory text or labels horizontally, from left to right, whenever possible. This does not apply to the y-axis of a line graph for data presentation. Otherwise, it creates awkward visuals.

5.4.2.5 Colors and shades

You may use colors and shades for highlighting or emphasis in figures and tables; however, use colors with caution. Colors might be problematic. For example, a red cross is often used as a medical sign in North America, but it represents danger and a green crescent signifies medical services in Muslim countries. On the contrary, red is generally a positive color in east Asia, especially the greater China areas. However, a red cross marked over an individual's name means that person is executed. Choose neutral colors for the figures when you need to color their elements.

Figure 5-1, for example, is part of a figure used in a dissertation that is written by a doctoral student of Chinese nationality. The student submitted his dissertation as part of the requirements of PhD degree at a Canadian university. The dissertation has many visuals produced with unnecessary red color.

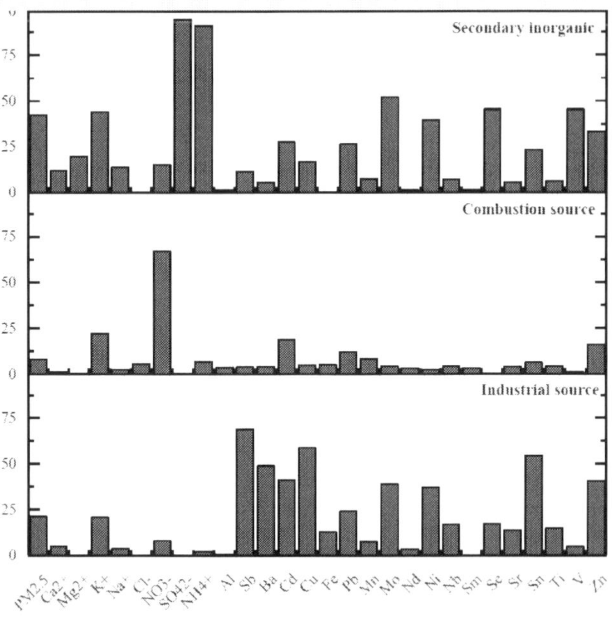

Figure 5-1. Part of a figure with unnecessary red color

Thoughtful writers think beyond their own cultures and experiences when they are writing for the international readers. Make sure that the graphics are free of religious or gender implications. If a human body or body parts are used in the visual, for example, it is recommended to represent the human beings with outlines or neutral abstractions – nudity is eccentric in engineering.

Black-and-white illustrations work well for most technical writing. You can differentiate them with different line styles, different thickness, and adding different dots. Avoid shades that are close to the background color. Otherwise, color mismatch creates negative visual effects and it may call for negative perceptions on the creator's capability and intelligence.

Figure 5-2 shows a line graph that is produced by using black elements only. Different line thickness and styles are used to differentiate them from each other. It achieves clarity without colors.

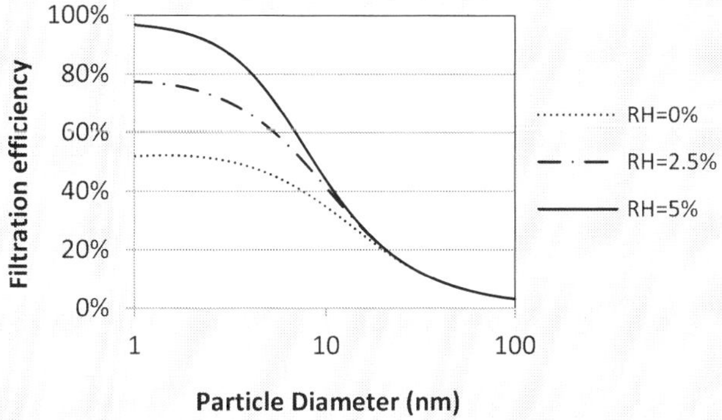

Figure 5-2. A black & white figure

In contrast, Figure 5-3 shows a line graph with errors. It is used in a Master's thesis (draft) collected in an American university. The author uses pink color (line and squares) for the data obtained at 37 °C and yellow color (line and triangles) for data collected at 52 °C. The yellow line and triangles are barely visible on the white background. This color problem can be fixed by using black lines and dots with different shapes and sizes (*see* Figure 5-2).

In addition to the unneeded colors, Figure 5-3 has several other errors as follows.

- The title of the graph ([H+]/[H+]0 *vs.* time of rice straw) is located at the top of the figure without emphasis.
- The y axis ([H+]/[H+]0) omits words and unit, leading to unclearness.
- The x axis name (time, hr) at the bottom is separated from its scale at the top.
- Equations and R-squares are separated from the data in graph (not clear what they are for)
- The legends collapse on the lower-right edge of the figure boarder.
- The inner board is open on the right side.
- Minor grid is used for y axis, but not for the x axis.

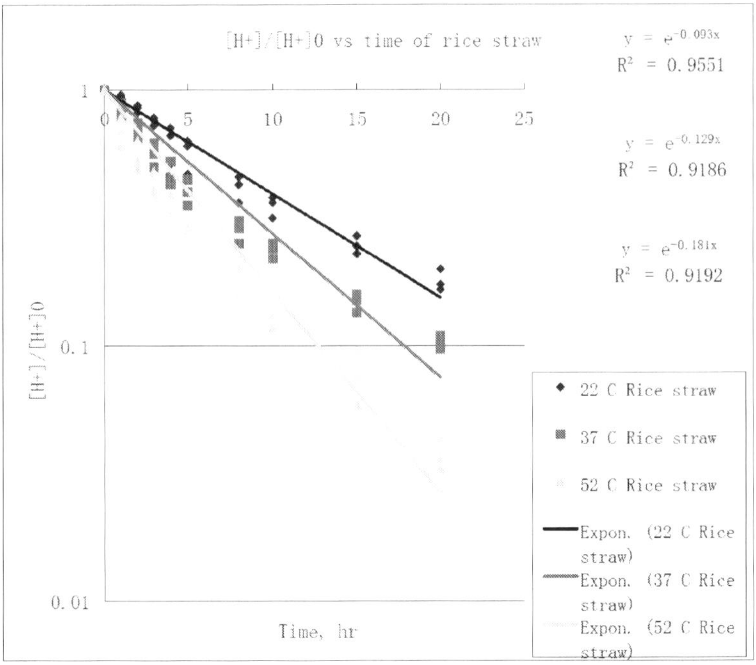

Figure 5-3. Figure with blunders

5.4.2.6 Graphs

A graph presents numerical data; it is more apparent and comprehensible than using a table for the same set of data. However, graphs are not as precise as tables; data in tables may contain multiple digits after decimal. You can create line graphs, bar graphs, pie graphs, *etc.* using data that can be presented in tables.

Line graphs (combined with dots) are usually used in engineering publications, and other types of graphs are primarily for non-technical presentations. A line graph allows the comparison between two or more sets of data (y-axis) against one common variable (x-axis). Figures 5-2 and 5-3 are examples of line graphs.

As needed, use appropriate scales in visuals for clarity. You should avoid wrong scales that may cause visual distortion and those that which may lead to inaccurate presentation. (*See also* Chapter 2) Logarithmic scales can help reduce misleading information in line graphs. A logarithmic scale enables displaying data over a wide range of measurement values in a compact way.

Figure 5-4 and Figure 5-5 show two images and a dotted graph, respectively (Li *et al.*, 2019). Pay attention to the scales, axis's, text directions in the visuals. They are important to the clarity of presentations. Reference scales must be used when size matters to your argument.

Figure 5-4. Photos used as visuals.

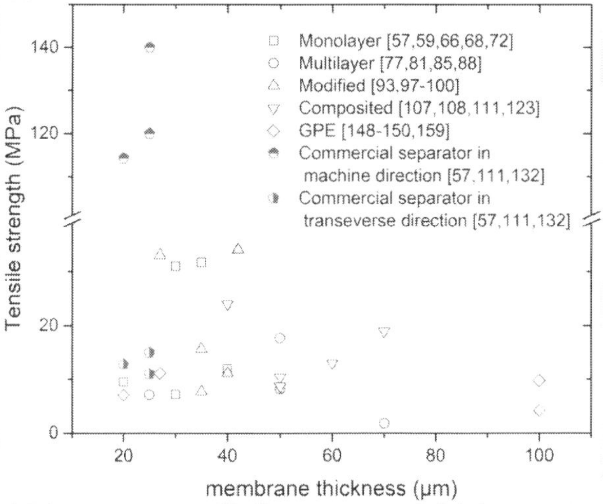

Figure 5-5. Dotted graph

5.4.2.7 Schematic diagrams

Schematic diagrams are frequently used as visuals in engineering publications. A schematic, or schematic diagram, is a type of visuals for illustration of the relationship among components of a system. A schematic is created by using abstract, graphic symbols and lines instead of realistic photos. Schematics focus on the emphases, details, and relationship that pictures cannot illustrate.

Schematics can be simple or complex. They are often used in engineering publications to depict complex systems (*e.g.* experimental setups) or processes. In the schematics, blocks and texts are used to indicate major devices and accessories, which are connected by lines. Arrows are used to indicate the direction of flow or order of sequence.

Figure 5-6 shows the schematic diagram of an experimental setup used for air filtration studies. Givehchi (2015) used this diagram to describe the procedure of air filtration tests. It is produced with black color only. It is plain in color, but clear enough.

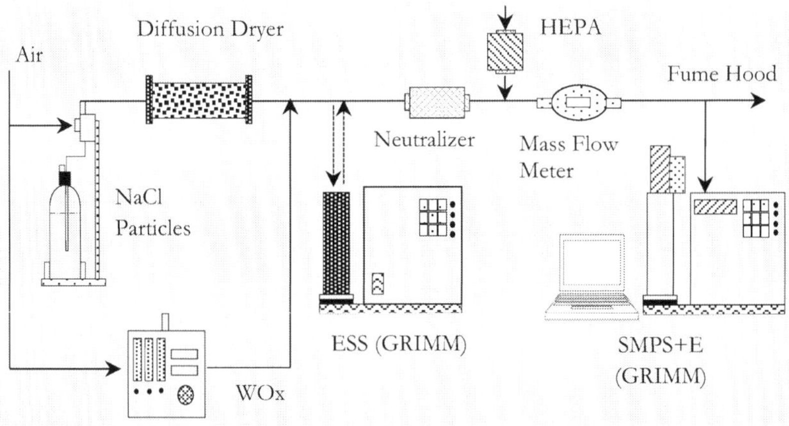

Figure 5-6. Schematic diagram of an experimental setup

Figure 5-7 shows a schematic drawing of an engine piston (Tan, 2014). Color is not applied to the lines and shapes in this engineering diagram, but the variation in lines and shapes improves clarity. Optional purple font is used to differentiate physical parts (cylinder, crank, piston, and rod) and the model parameters (angle, bore, radius, and stroke), which are defined to assist engineering analyses. Shaded piston stands out from the rest at the top because of the contrast effect. Labels are aligned on the left or right with consistence, and white spaces purposely surrounding the elements further enhance visual comfort. Overall, this is a well-prepared visual.

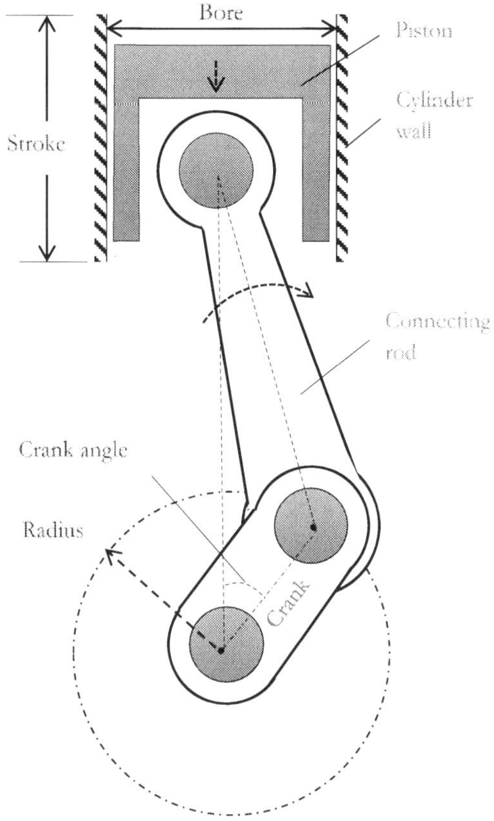

Figure 5-7. Schematic diagram of a four-stroke piston engine

5.4.2.8 Exploded view

Exploded view is another type of schematics. An exploded-view schematic reveals the proper sequence of assembly or the details of individual parts. It is also useful to the illustration of internal parts and their relationship to the whole device or system. Exposed view graphs appear frequently in manuals; however, it is rarely used in engineering academic publications.

5.4.2.9 Maps

Maps are useful to the presentation of geographic information. Depending on the purpose of the writing, a map may have the following information: climate patterns, distributions of populations, geographic roads, location of sites, traffic, and so forth.

Follow the general guidelines about visuals for the creation and integration of maps into the text. In addition, you need to include a scale for the indication of distance and the direction of north for orientation. You can emphasize key items by using color, shading, and so forth. On the other hand, make sure you eliminate unnecessary information to avoid clutter.

5.5 Tables

Tables are more effective than graphs for the *precise* presentation of data used to support your reasoning and argument. Modern word processing software has a variety of choices for the creation of tables. The following is only to emphasize clarity and conciseness in writing when you use tables.

The table contents should be descriptive, but brief. For complex tables, use abbreviations and symbols in box head. Footnotes can be used for defining abbreviations or symbols if they are not defined in the text of the main document. Within the table body, align the data in adjacent rows or columns for the ease of comparison.

Always attempt to fit a table on one single page. Otherwise, repeat the table number of the same table on the new page. The caption title usually does not repeat on the new page. If it repeats, the caption title should be identical to that on the previous page, but it is followed by a comma mark and a word *continued,* or the word *continued* enclosed in parenthesis. For example, Table 3. continued, Table 3. (continued), Table 3. [Caption title], continued, or Table 3. [Caption title] (continued).

Table 5-2 shows how one table is presented on two pages in the printed article by Li *et al.* (2019). The example tables are reproduced for simplicity. The caption title is omitted on the new page. The column headings (and stub headings, column spanner, *etc.*) are duplicated from the previous page, even though there is only one cell on the new page.

Table 5-2. A table presented on two pages

Table #. Monolayer electrospun membrane separators for LIB

Materials	Fiber diameter	Liquid electrolyte	Electrolyte uptake	Thermal dimensional stability	…
PVDF	-	$LiPF_6$-EC/DMC	~400 %	-	…
	…	*[multiple rows omitted]*	…		
PI	500 nm	LiBOB-PC	270-340 %	Unchanged at 150°C for 1h	…

Table #. (continued)

Materials	Fiber diameter	Liquid electrolyte	Electrolyte uptake	Thermal dimensional stability	…
				Unchanged at 500 °C for 2h	…

5.6 Equations

You often need chemical, physical, and mathematical equations to improve clarity in technical writing. Write the equations with symbols and Arabic numbers. The symbols can be defined and listed in the list of symbols. You can also define the symbols and list them immediately after the equation where they are first used.

Regardless of the type of equations, each equation takes at least one line by itself. Unlike visuals, *Table of Equations* is not needed anywhere in the document. Table 5-1 shows the cross-reference formats of equations.

Avoid words in equations in academic technical writing. Example 5-10 shows the word equations written into the theses of two international students, who studied in north American universities.

This example emphasizes the importance of Arabic numbers and symbols to the conciseness of writing.

Example 5-10. Chemical and mathematical equations

Incorrect: Hydrogen gas + oxygen gas → steam (5-1)

Revised: $2H_2(g) + O_2(g) \rightarrow 2H_2O(g)$ (5-2)

where the letter g in parentheses stands for *gas*.

Incorrect: $Force = Mass \times Acceleration$ (5-3)

Revised: $F = m \times a$ (5-4)

where F = force (N), m =mass (kg), and a = acceleration (m/s^2)

Incorrect:

$$\text{Uncertainty} = \sqrt{(Error\ fraction\ \times concentration)^2 + (0.5\ \times MDL)^2} \quad (4.3)$$

Revised:

$$\alpha = \sqrt{(fC)^2 + (0.5MDL)^2} \quad (4.3)$$

where α is the uncertainty, f is the error fraction, C is the concentration, and MDL is the method detection limit.

~~~

After editing the paragraphs, visuals, and equations following the guidelines in this chapter, your next revisions should focus on the sentences. You may have done so while you are editing the paragraphs, but the next chapter, *Sentences*, helps you further improve the quality of your manuscript.
~~~

6 Sentences

6.1 Sentence Construction Basics

Sentences are constructed with words, phrases, clauses, and other elements following the writing rules and structural styles. The *subject-verb-object* pattern is the most common sentence structure in English. Every sentence, except commands, must have a subject and a verb. A sentence using the active voice should have a subject. The verb and its subject should be in grammatical agreement. Expletives (*e.g., there is*; *it*) can be used to move the subjects away from their normal positions in the sentences.

There are a variety of other sentence structures to improve clarity and conciseness in writing. These advanced topics are introduced in Sections 6.2 and 6.4. after the basics are selectively introduced in this section.

6.1.1 Grammatical agreement

Grammatical agreement is a form of correspondence among subject, verb, object, and other elements of a sentence. For instance, the subject, verb, and number in the same sentence must be in grammatical agreement; pronouns must agree in *person*, *number*, and *gender*.

6.1.1.1 Subject-verb agreement

Pay attention to singular and plural pronouns. *One, each, series* and *portion* are singular nouns; occasionally they precede plural nouns. Indefinite pronouns (*all*, *more*, *none*, and *some*), either in their singular or plural forms, can be used with mass nouns or count nouns, respectively. Relative pronouns (*that*, *which, who, whom, whose*) can precede either singular verbs or plural verbs, depending on the context.

Example 6-1. Subject-verb agreement (1)

One out of ten truck drivers interviewed at the rest areas is female. [*The verb is must agree with one instead of truck drivers or rest areas.*]

Some of the carbon dioxide has become supercritical fluid [*Carbon dioxide* is a mass noun.]

Most of the truck drivers do not know that air in the truck cabin is polluted [*drivers* is plural].

A large portion of journal articles in this field is devoted to the relationship between climate change and the solar radiation.

The reading of the pressure gage, which is recorded by the computer, is 20 kPa.

Pay attention to the temperatures of the samples that are monitored every minute.

Not all words ending with *-s* are plural subjects. For example, *analysis* is singular, and its plural form is *analyses.* Abstract nouns (*ethics, mathematics, news, physics, etc.*) appear to be plural but are actually singular. Furthermore, a non-native writer in English may want to pay close attention to the following confusing cases.

1) Verbs must agree with subjects regardless of the number of a subjective complement.

Example 6-2. Subject-verb agreement (2)

Wrong: The topic of this report are nanomaterials.
Right: The topic of this report is nanomaterials. [*The word nanomaterials may be distracting.*]

2) Subjects that expresses measurement, weight, mass, or the like take singular verbs.

Example 6-3. Subject-verb agreement (3)

Wrong: At least three numbers are needed to calculate the standard deviation of those numbers.
Right: At least three numbers is needed to calculate the standard deviation of those numbers.

3) Some words, like *scissors*, are plural only.

Example 6-4. Subject-verb agreement (4)

A pair of scissors was used to remove the string.
Scissors were used to remove the string.

Regardless of singular of plural, the title of a long document like a book and a thesis takes a single verb in a sentence.

Example 6-5. Subject-verb agreement (5)

Wrong: *The Lord of the Rings* are an epic high-fantasy novel.
Right: *The Lord of the Rings* is an epic high-fantasy novel.

6.1.1.2 Compound subject-verb agreement

Compound subjects are joined by a conjunction (Section 6.3.1). Two parts joined by *and* normally take a plural verb (Example 6-6). However, there are two exceptions to this grammatical rule.

Exception 1. When the compound subject is generally thought to be a single unit. *See* Example 6-7.

Exception 2. A compound subject becomes singular when it is modified by each or every. *See* Example 6-8.

Example 6-6. Compound subject-verb agreement (1)

Wrong: The severity and frequency of these events has introduced the conversation of global warming and climate change to the forefront of many political debates.
Right: The severity and frequency of these events have introduced the conversation of global warming and climate change to the forefront of many political debates.

Example 6-7. Compound subject-verb agreement (2)

Wrong: Peanut butter and jelly are my favorite.
Right: Peanut butter and jelly is my favorite.
Wrong: Two and three equal five.
Right: Two and three equals five.

Example 6-8. Compound subject-verb agreement (3)

Wrong: Each conclusion and recommendation are important to the readers.
Right: Each conclusion and recommendation is important to the readers.
Wrong: Each figure and table have to be numbered.
Right: Each figure and table has to be numbered.

A compound subject can be made up of one singular element and one plural element that are joined by *or* or *nor*. This type of subjects requires the verb to agree with the closer element.

Example 6-9. Compound subject-verb agreement (4)

Wrong: (Either) Table 1 or Figures 4-6 supports the argument.
Right: (Either) Table 1 or Figures 4-6 support the argument.
Right: (Either) Figures 4-6 or Table 1 supports the argument.

Note:

Non-native English writers may split the compound subject and use two sentences with simple subjects when you are not sure whether the verb is plural or singular.

6.1.1.3 Grammatical agreement considering gender inclusivity

Personal pronouns must agree in gender with their antecedents, which are the nouns they refer to. For plural pronouns, agreement in gender is not an issue because they automatically agree with antecedents. For singular, gender may be a challenge to non-native English writers because, in some languages (*e.g.*, Chinese), singular pronouns are not gender specific. In English, the singular pronouns (*he/him/his*, *she/her/hers*, and *it/its*) are gender specific, and you need to match them with the genders of their antecedent.

The existence of other genders has been recognized among some nations and religions for centuries. With the growing awareness of gender inclusivity, gender consists of not only male and female, even in traditional English-speaking countries. Like it or not, several countries have passed laws recognizing sex identities other than the male or female bodies.

You can consider the following approaches to gender inclusivity in your writing.

1) Without sacrificing clarity, use indefinite pronouns (*everyone, anyone, person, individual, each, etc.*) and nouns (*member, student, performer, child, person, etc.*) that do not distinguish males or females.
2) When you have to, use *he or she*, *him or her*, and *his or her*, especially when the gender is not identified.

Example 6-10. Gender inclusivity (1)

He or she introduced the speaker to the audience.
Every author has provided his or her email address.

The combinations of *he or she*, *his or her*, and so on impacts the smooth flow of reading if repeated in one sentence. Alternatively, you can take any of the following two options.

3) Use *they/their/theirs* with a plural antecedent
4) Rewrite the sentence to remove the pronoun

Example 6-11. Gender inclusivity (2)

All authors have provided their email addresses.
The moderator introduced the speaker to the audience. *[rewording]*

Note:

Avoid gender-related bias. For example, using "*the nurse . . . she...*" or "*the engineer . . . he...*" without confirmation is not acceptable in English writing. Both male and female can be nurses or engineers in many countries.

6.1.2 Clauses

6.1.2.1 Dependent and independent clauses

A clause can be a dependent or an independent clause. An independent clause stands alone as a simple sentence. Otherwise, it is a *dependent* clause or subordinate clause. Each sentence must contain one or more independent clauses, but dependent clauses are used for effective expression of thoughts and establishment of their relative importance.

Example 6-12. Clause importance

Equal importance:

The landfill site in Waterloo region is in the City of Waterloo. It is used by the cities of Cambridge, Kitchener, and Waterloo.

Subordinated:

The landfill site in Waterloo region, which is in the City of Waterloo *[dependent]*, is used by the cities of Cambridge, Kitchener, and Waterloo.

6.1.2.2 Adjective clauses

A noun precedes an adjective clause. Adjective clauses can be particularly tricky for non-native writers of English because clauses are formed in a variety of ways in different languages.

Example 6-13. Adjective clauses

Wrong: The one who she is giving the presentation is my student.

Right: The one who ~~she~~ is giving the presentation is my student.

Wrong: The student is giving a presentation who is standing in front of the screen.

Right: The student who is standing in front of the screen is giving a presentation.

Comments:

The adjective clause *who is standing in front of the screen* modifies *student instead presentation.* Therefore, it must immediately follow the word *student.*

6.1.3 Modifiers in sentences

In short, a modifier is a describer. It can be a word, a phrase, or a clause. You do not need modifiers to construct grammatically correct sentences, but you often need modifiers to provide details and to reinforce clarity. It is especially important to technical writing because many modifiers (words, phrases, or clauses) transform general elements into more specific ones in sentences.

Example 6-14. Modifiers in sentences

Unmodified: Sizes increased with potential. *[what size? What potential?]*

Modified: Nanofiber sizes increased with the potential at the needle tip.

Most modifiers function as adjectives that impose constraints on the words they modify. For instance, *ten* replications, *this* figure, *that* conclusion, *etc.*

6.1.3.1 Prepositional phrases as modifiers

A prepositional phrase can act as a modifier. It is normally composed of the preposition, its object, and the object to modify. Be cautious with awkward prepositions in sentence construction.

Example 6-15. Prepositional phrases as modifiers

You should write an article following the ethics guideline. *[The prepositional phrase following the ethics guideline modifies the verb write.]*

The results ~~are where they~~ were presented above.

6.1.3.2 Restrictive and non-restrictive modifiers

Phrase and clause modifiers may be restrictive or non-restrictive. A *restrictive* modifier restricts the meaning of the element it modifies; otherwise, the modifier becomes non-restrictive. Omission of non-restrictive modifiers does not affect the main idea of the sentence.

The same sentence may have different meanings, depending on whether the modifier is restrictive or non-restrictive. Pay close attention to the commas that sets off the modifier to avoid misleading sentences.

Example 6-16. Non-restrictive modifier

The coronavirus disease 2019 (COVID-19), which was officially named as severe acute respiratory syndrome coronavirus (SARS-CoV-2), quickly spread around the world in early 2020.

Comments:

This sentence has the same meaning as *The coronavirus disease 2019 quickly spread around the world in early 2020.* The non-restrictive modifier sets off by commas provides non-essential but extra information.

Note: which/that

In American English, *that* is used to introduce restrictive clauses, and *which* introduces non-restrictive clauses. In British English, *which* (not *that*) can introduce in restrictive clauses too. A non-native writer in English can always use *which* to introduce non-restrictive clauses but use *that* to introduce restrictive clauses. It helps avoid confusions to the readers who are only familiar with American English.

6.2 Sentence Structure Types

A sentence structure can be simple, compound, complex, or compound-complex. An independent clause forms a *simple sentence*; multiple independent clauses form a *compound sentence*. Commas, semicolons, or coordinating conjunctions are needed in a compound sentence for clarity. (*See* Table 6-1. Conjunctions). A *complex sentence* contains a subordinate clause or clauses. A sentence that has multiple independent clauses and one or more dependent clauses is referred to as a *compound-complex sentence*.

Example 6-17. Sentence structure types

Simple sentences:

Summer is here.

I am writing a book.

Compound sentences:

Aerosol samples can be collected using air filters for morphology analysis, but it is not the only way to characterize aerosol particles.

The sizes of nanoparticles are in the range of 1-100 nm; various chemicals are present in the particles at different proportions.

Complex sentence:

The power to the corona charger will shut off automatically *[independent clause] when* the door is open *[dependent clause]*.

Compound-complex sentence:

Fiber size is important to the filtration efficiency of a filter *[independent clause]*; the filtration efficiency drops *[independent clause]* when the fiber size increases *[dependent clause]*.

Both long and short sentences are needed in academic writing as long as their meanings are clear to the readers. Uncomplex sentences are normally used in technical writing for the presentation of complex ideas. However, nothing is wrong with complex sentences; they are useful to explanations and convincing arguments. Over complex sentences, however, are likely to confuse the readers whose non-native language is not English. Simple sentences are easy to follow, and short sentences are generally used for direct statements.

Experienced writers use various sentences in different lengths, structures, and complexities to achieve lively effects. A series of sentences of the same style becomes tedious and monotonous. Consequently, it loses the attention of the readers. Varying sentence styles can create effective contrast and emphasis (*See also* 6.3.3.1 and 6.4.2.)

6.3 Constructing Effective Sentences

Parallel structures and conjunctions are commonly used for the construction of effective sentences. They are used to achieve clarity, conciseness, emphasis, and variety. These writing techniques are selectively explained as follows.

6.3.1 Parallel sentence structures

Parallel structures are developed with sentences having the same grammatical structure. That is, adjectives are paralleled by adjectives, nouns by nouns, verbs by verbs, and so forth. Parallel structures can be constructed by using words, phrases, clauses, and sentences, but they should not be mixed.

Consistence is the key to effective parallel structures. In contrast, inconsistence can confuse your readers and weaken your argument. As shown in Example 6-19, words and clauses should not be used in parallel.

Example 6-18. Parallel structures in sentences

(1) **Parallel structure in preverbs:**

Easy come, easy go.

No pain, no gain.

Parallel structure created by two or more sentences:

"It is by logic we prove, but by institution we discover" (Leonardo DaVinci)

"Humanity has advanced, when it has advanced, not because it has been sober, responsible, and cautious, but because it has been playful, rebellious, and immature." (Tom Robbins, *Still Life with Woodpecker,* 1980)

Example 6-19. Parallel structures in sentences (2)

Unparallel: This research project is interesting and a challenge. *[Adjective by noun.]*

Parallel: This research project is interesting and challenging. *[Adjective by adjective.]*

Unparallel: A sophisticated air quality monitor may be used to detect these air pollutants: sulfur dioxide, volatile organic compounds, ozone, and *measuring* particles.

Parallel: A sophisticated air quality monitor may be used to detect these air pollutants: sulfur dioxide, volatile organic compound, ozone, and ~~measuring~~ particles.

Unparallel: Wildfires are more intense, droughts are commonplace, and in humid climates hurricanes are more and more expected.

Parallel: Wildfires are more intense, droughts are more common~~place~~, and hurricanes are more expected than before in humid climates.

Unparallel: In typical scholarly writing, the authors are expected that they would start with background information, that they would elaborate on the methodology, and that results would be presented to the readers. *[The tone shift from positive to passive breaks the parallel structure.]*

Parallel: In typical scholarly writing, the authors are expected that they would start with background information, that they would elaborate on the methodology, and that they would present the results to the readers.

6.3.2 Sentences by conjunctions

A conjunction connects words, phrases, or clauses in a sentence to explain the relationship between these elements. As seen in Table 6-1, there are four types of conjunctions: coordinating conjunctions, correlative conjunctions, subordinating conjunctions, and conjunctive adverbs. They are used in this book, and many other engineering publications.

Subordination is the act of giving someone or something less importance or power. In technical writing, specifically, subordination is the process of subordinating secondary ideas in subordinate elements. Effective subordination is useful to achieving conciseness, emphasis, and sentence variety in writing.

Table 6-1. Conjunctions

Conjunction	Example	Function
Coordinating conjunction	and, but, or, for, nor, yet, so…	Joins two sentence elements with identical functions
Correlative conjunctions	either/or, neither/ nor, not only/but also, both/and, whether/or, not/but…	Used in pairs (*See* Parallel structure)
Subordinating conjunction	after, although, as, as if, as long as, because, before, despite, even if, even though, if, in order that, rather than, since, so that, that, though, unless, until, when, where, whereas, whether, while…	Connects the elements of a sentence and distinguishes their relative importance
Conjunctive adverb	accordingly, besides, consequently, for example, for instance, in another word, further, however, moreover, nevertheless, on the other hand, then, therefore…	Joins two independent clauses

6.3.2.1 Correlative conjunctions

As suggested by their names, correlative conjunctions correlate and work in pairs to join words, phrases or clauses in a parallel structure. The elements (words, phrases, or clauses) in both parts of the pairs should have similarity and follow the same grammatical form.

Example 6-20. Correlative conjunctions

Parallel words:

Corona viruses carry either DNA or RNA, never both.

Parallel phrases:

Both the nurses and the patients are exposed to the viruses.

Parallel clauses:

Either we work together to fight the pandemic, or we let the pandemic beat us all.

Your technical proposal not only must have a list of objectives, but also should include a list of deliverables.

6.3.2.2 Subordinating conjunctions

Subordination is effective in distinguishing the relative importance of sentence elements. Be careful with the main idea of the sentence when you use grammatically subordinating elements. The main idea of a sentence depends on the position of the subordinating element. Both sentences in Example 6-21 are logical, but their main ideas differ from each other.

Example 6-21. Subordinating conjunctions (although)

Although cobalt based solvents are effective in absorbing NOx, they create secondary pollution. *[The main idea is effective cobalt based solvents.]*

Cobalt based solvents create secondary environmental pollution, although they are effective in absorbing nitric oxide. *[The focus of this sentence is cobalt based solvent create secondary environmental pollution.]*

Example 6-22. Subordinating conjunctions (because)

Providing excuses:

Because the relative humidity level in the laboratory is not controlled properly and the de-humidifier in the testing compartment failed, the results contain artifacts.

Providing facts:

The results contain artifacts because the relative humidity level in the laboratory is not controlled properly and the de-humidifier in the testing compartment failed.

6.3.3 Sentence variety

Monotonous writing is boring to your readers, and yourself too. You can write interesting sentences with modifiers (*See* 6.1.3 and 6.5.5). On the other hand, overuse of modifiers can be monotonous too. The best approach is to use a variety of sentences.

There are several techniques to achieve sentence variety, including varying sentence length and inverting word order. They are briefly introduced as follows.

6.3.3.1 Varying sentence length

Connect short independent clauses by subordinations (*See* Table 6-1). You can also combine short sentences into a long one by converting verbs into adjectives.

Example 6-23. Sentence variety (length and tone)

Monotone:

The distillation column is 20 meters tall, and its diameter is 4 meters, and its surface roughness is 5 micrometers.

Revised:

The distillation column, which is 20 m tall and 4 m in diameter, has a surface roughness of 5 micrometers.

6.3.3.2 Varying word order

Inverting the word order or inserting a phrase or a clause enhances sentence variety. This technique is effective in achieving emphasis, providing details, and controlling pace.

Example 6-24. Sentence variety (Word order)

Inverting word order:

Normal: The photo of blackhole has never been so clear.

Inverted: Never has the photo of blackhole been so clear.

Inserting elements:

A high-speed air flow creates friction on the surfaces, both top and bottom, of the plate.

6.4 Emphasis in Sentences and Paragraphs

Emphasis is essential to clarity by stressing the primary ideas of sentences and paragraphs. Effective emphasis can be accomplished by alternating position, using repetition, using intensifiers, varying sentence length, changing sentence type, and so forth. Using special fonts like *italics,* **bold** face, underlining, and CAPITAL letters also enhances emphasis. However, use them occasionally to avoid visual clutter. Another common approach is to use phrases like *again*, *most importantly*, and *foremost.* You can also use dash marks for similar purposes (*See* 8.5. Dashes).

6.4.1 Emphasis by position

Although the *subject-verb-object* pattern is most familiar to English writers, you may occasionally construct an inverted sentence to achieve emphasis on your main ideas. You can catch the readers' attention by placing the elements in an unusual order to achieve emphasis.

The beginning and end elements of a sentence, a paragraph, or a document catch readers' eyes. Therefore, start a sentence, a paragraph or a document with the ideas that you wish to emphasize.

Example 6-25. Inverted sentence

Normal: Uncomplex sentences are preferred in technical writing for the presentation of complex ideas. *[Emphasize the value of Uncomplex sentences.]*

Inverted: In technical writing, Uncomplex sentences are preferred for the presentation of complex ideas. *[Emphasize technical writing; it may not be true for another type of writing.]*

Another way to emphasize by using position is using lists. You can list your ideas allowing the readers to understand their order of importance in sequence (*See* Lists).

A typical mistake that non-native English writers make is preceding the main idea of a sentence with less important elements. Non-native English writers, for instance, often put the purpose, location, reason, *etc.* before the main idea of a sentence; For example, most Chinese consider being straightforward as being impolite. Even in their verbal communications, they respond to an inquiry beginning with reasons. These indirect introductory elements, however, will demote the importance of the main idea and confuse the readers.

Example 6-26. Emphasis by positions of elements in sentences

Incorrect: Inside the fume hood, the space is filled with materials and supplies.

Correct: The space inside the fume hood is filled with materials and supplies.

Incorrect: In order to establish a quantitative relationship between winter road condition and vehicular air emissions, multiple linear regression models were developed (Min, 2015). [*The scope of the thesis work is models.*]

Correct: Multiple linear regression models were developed to establish a quantitative relationship between winter road condition and vehicular air emissions

Incorrect: If the process can be improved, solar fuels would provide many benefits for society. *[by a native speaker of English]*

Correct: Solar fuels would provide many benefits to society if the process can be improved.

Incorrect: Another factor affecting the clarity of writing is point of view.

Correct: Point of view is another factor affecting the clarity of writing. *[This is the sentence under heading 4.2.6]*

6.4.2 Emphasis by sentence length

Strategically varying sentence length serves the purpose of emphasis in a paragraph. A short sentence that follows a long one, for example, usually catches the readers' attention.

Example 6-27. Emphasis by a short sentence

We have conducted a start-of-the-art literature review on the methodology and key findings of nanofiber fabrication technologies. All lead to a conclusion that, regardless of the advances that were made recent years in nanomaterials, nanofiber by electrospinning is a relatively new field of research. More research is needed.

6.4.3 Emphasis by repetition

Repeating a keyword, a sentence, or a paragraph emphasizes an idea in a powerful way. In engineering writing, it is better to repeat consistently the elements than to use synonyms. On the other hand, aimless repetition may impact the conciseness because the sentence or paragraph might appear awkward and pointless. Therefore, use repetition technique with caution.

Example 6-28. Emphasis by element repetition in sentences

Poor repetition:

Recent reports concur our conclusions in this paper. These studies confirm that it is feasible to reduce greenhouse gas emissions by our technology. Their analyses, however, are primarily computational models. [*Changing words from reports to studies to analyses might help with variety, but not for emphasis*. These three words are slightly different in meaning: analyses are part of the studies that lead to the reports]

Effective repetition:

Recent studies concur our conclusions in this paper. These studies confirm that it is feasible to reduce greenhouse gas emissions by our technology. Theirs, however, are primarily computational models.

6.4.4 Emphasis by intensifiers

The adverbs that emphasize degree are called intensifiers; for example, *very, rather, much,* and *too* are intensifiers. They serve a necessary function of comparison but use them with caution in technical writing. Some intensifiers (such as *perfect, impossible,* and *final*) cannot be used for intensification because scientific research never stops advancing. Overuse qualitative intensifiers (*very fine, too high, much longer*) may also cause vagueness and inaccuracy.

Example 6-29. Replacing vague intensifiers with values

Vague: Our results agree very well with theirs.

Clear: The difference between our results and theirs is less than five percent.

6.5 Sentence Faults

Typical sentence faults include rambling sentences, fragmented sentences, illogical assertions, and overuse of modifiers. The following highlights the sentence errors that non-native English writers may make.

6.5.1 Garbled sentences

Garbled sentences lose emphasis because all ideas are presented with equal importance. Avoid loading your sentences with multiple thoughts.

Example 6-30. Garbled sentences

Garbled: The company was founded in 2019, only three members were on the staff, and all members took multiple positions and roles, but it was not an efficient operation.

Revised: When the company was founded in 2019, only three members were on the staff, and all members took multiple positions and roles. However, it was not an efficient operation.

6.5.2 Rambling sentences

Avoid rambling sentences because they overload the readers with more information than the readers can comfortably take. You can avoid this problem by dividing a rambling sentence into multiple sentences, starting with one that delivers your main ideas.

Example 6-31. Rambling sentences

Rambling sentences:

> In a CSTR, four dispersion patterns (flooding, cavity formation, complete dispersion of gas, recirculation of gas-liquid mixture) can be observed and the minimum impeller speed for complete dispersion of gas phase, N_{cd} is a key parameter considering the fact that gas-liquid contacting at impeller speeds below N_{cd} leads to a poor reactor performance as the lower part of reactor is wasted due to the incomplete gas dispersion. ……

Revisions to reduce rambling

> There are four dispersion patterns, flooding, cavity formation, complete dispersion of gas, and recirculation of gas-liquid mixture, in a CSTR. A minimum impeller speed is needed for a complete dispersion of gas phase. A lower speed leads to incomplete gas dispersion at the lower part of the reactor.

6.5.3 Fragmented sentences

A sentence becomes fragments when it omits key sentence elements, such as a subject or a verb. A stand-alone subordinate clause or phrase is also a sentence fragment. Sentence fragments can occur not only in long complex/compound sentences, but also in simple ones.

6.5.3.1 Missing verbs

Missing verbs causes fragmented sentences. A sentence must have at least one verb. It is not a problem when the sentence is short and simple, but missing a verb is. It may occur unintentionally, especially when the sentences become complex. In addition, verbals cannot replace verbs for the same grammatical functions. (*See* 7.4.3)

Example 6-32. Sentence fragment (missing verbs)

Fragment: And record the reading displayed on the screen.
Revision: He records the reading displayed on the screen.
Fragment: Supporting your argument with examples.
Sentence: You need to support your argument with examples.

6.5.3.2 Missing subjects

Missing subjects occurs when the writers assume them to be something by default, leaving the readers puzzled.

Example 6-33. Sentence fragments missing subjects (1)

Fragment: The increasing carbon dioxide concentration in the atmosphere causes to rise over the last century.
Sentence: The increasing carbon dioxide concentration in the atmosphere causes the ocean temperature to rise over the last century.

6.5.3.3 Other sentence fragments

Sentence fragments often result from the usage of relative pronouns (*e.g. who, which*, and *that*) or subordinating conjunctions (such as *although, because, if*, and *when*). The usage of explanatory phrases (*such as, for example*) may also create sentence fragments.

Example 6-34. Sentence fragments missing subjects (2)

Fragment: The biological approach comes with some benefits. For instance, contributes to environmental sustainability.
Sentence: The biological approach comes with some benefits. For instance, it contributes to environmental sustainability.

6.5.4 Illogical assertions

Illogical assertions cause faulty logics. Obvious illogical assertions can be avoided, but they may be tricky, especially to non-native English writers.

Example 6-35. Illogical assertions in sentences

Illogical: The probe's length is 20 cm long. *[Duplicated function]*
Logical: The probe is 20 cm long.
Logical: The probe's length is 20 cm.

6.5.5 Modifier errors

6.5.5.1 Overuse of modifiers

Overuse of subjective modifiers such as *surprisingly*, *absolutely*, *obviously*, *unfortunately*, and *definitely* implies emotion and undermines the logics of argument. These types of modifiers impose your personal feelings or opinions on the readers without concrete evidence. (*See also* 4.2.7)

Example 6-36. Avoid emotional writing

Incorrect: Surprisingly, they did not report the error analysis in their publication.
Correct: ~~Surprisingly ,~~They did not report the error analysis in their publication.

Incorrect: We are definitely certain that it was the catalyst fouling which caused the damage.
Correct: We are ~~definitely~~ certain that it was the catalyst fouling that caused the damage.

Incorrect: Obviously, clean air is absolutely necessary for a healthy and productive economy. *[by a native English speaker]*
Correct: ~~Obviously,~~ Clean air is ~~absolutely~~ necessary for a healthy and productive economy.

Comments:

The adverbs *surprisingly, sadly, obviously* and *definitely* modify the entire clauses. Your readers may not feel the same way as you do.

Example 4-35. (*Reused here to illustrate emotional writing*)

… The quantitative analyses showed that the different drag models led to significant differences in dense phase simulations. Among the different drag models discussed, the Gidaspow model gave the best agreement with experimental observation both qualitatively and quantitatively. The present investigation showed that drag models had critical and subtle impacts on the CFD predictions of dense gas–solids two-phase systems such as encountered in spouted beds. *[Source: Du et al. 2006]*

Comments:

The vague words, significant, best, critical, and subtle are misused as intensifiers in this paragraph. This abstract becomes useless to the readers because it lacks supporting evidence. More about vague words are introduced in Section 7.2.2.

6.5.5.2 Misplaced modifiers

Misplaced modifiers cause ambiguity. A modifier should be placed immediately before or after the sentence element to be modified. This might be challenging to non-native English speakers. Note the shift of meanings in Example 6-37.

Example 6-37. Placement of modifiers in sentences

Words as modifiers:

He almost lost all the data collected last night. [*almost modifies lost; the data were not lost.*]

He lost almost all the data collected last night. [*almost modifies all; some data were lost.*]

Clauses as modifiers:

We sent the samples to the lab that we are satisfied with for analysis.

We sent the samples that we are satisfied with to the lab for analysis.

6.5.5.3 Squinting modifiers

Another type of misplaced modifiers is squinting modifier. A squinting modifier is out of place because it could modify two elements simultaneously. A squinting modifier causes ambiguity and confusion. Squinting modifiers can be corrected simply by rewriting the sentences.

Example 6-38. Squinting modifiers in sentences

Squinting: They decide immediately to start the investigation tomorrow. *[The word immediately could modify "decide" and "start the investigation."]*

Clear: They immediately decide to start the investigation.

Clear: They decide that they would immediately start the investigation.

6.5.5.4 Dangling modifiers

A dangling modifier has nothing to modify; it typically appears at the beginning or the end of the sentence. Dangling modifiers occur when writers get ahead of themselves. They assume that the context is obvious to the readers and forget to help the readers get ready with the context.

Like squinting modifiers, dangling modifiers confuses the readers, but for different reasons. A dangling modifier can be corrected by adding the appropriate subject to the modifier or the clause.

Example 6-39. Dangling modifiers in sentences

Dangling: While working on the lab report, the computer suddenly shut down *[It is not clear who is working. The computer cannot be working on a report and the writer assumed the reader to know that it is "I".]*

Correct: While I was working on the lab report, the computer suddenly shut down.

Dangling: The figure becomes more readable after replacing the symbols with descriptive words. *[Who replaced the symbols?]*

Correct: The figure becomes more readable when you replace the symbols with descriptive words.

Correct: The figure becomes more readable when the symbols were replaced with descriptive words.

~~~
~~~

7 Words and Phrases

"The difference between the almost right word and the right word is the difference between 'the lightning' and 'the lightning bug.'" - Mark Twain

7.1 Words for Communication Globally

The key to effective communication globally is recognizing the differences in cultures and languages. In a globalized world, it may become a challenge to the writer from one culture to communicate effectively with the readers from other cultures. What is considered efficient or acceptable in one language might be vague or blunt in another, and *vice versa.*

Writing for international readers posts a challenge to you, but it also offers you the opportunities to increase the impact of your publication. Your scholarly works can reach the readers worldwide. Therefore, it is important to choose commonly used words that are accepted by as many people as possible.

Precise wording is a basic requirement of clear technical writing. It is essential to learn the precise meanings and functions of the words. In addition to reading this book, you are encouraged to read one or two handbooks for formal writing (*e.g.,* Alred *et al.*, 2018). Understanding the grammatical terms like *antonym* and *synonym*, for example, allows you to choose the right words. Antonyms are pairs of words with meanings that are opposite to each other (*good/bad, interesting/boring, long/short, expensive/economical*). Synonyms are words of the same language that have the same or similar meanings. Examples include *average/mean, biannual/biennial, notorious/infamous, etc.*

A key to choosing the correct words is to keep growing your vocabulary. Reading, listening and daily conversation in English can help improve your vocabulary and techniques for writing.

International students, for example, are encouraged to read at least one article a day, and as much as possible. The reading materials should not be limited to the technical documents; reading newspapers, magazines, biographies, *etc.,* helps you in building vocabulary. In addition, watching television after work and listening to radios while driving are beneficial to vocabulary development.

Last, but not the least, practise writing as much as possible by using the words, especially in a technical context. It takes time and continual effort to become an effective writer. All these are useful to advancing your writing skills.

7.2 Word Errors

There are many ways to teach how to choose the right words. Instead of pretending to be rigorous like a handbook, this book is aimed to be practical and simple for technical writing. Many of us learn fast and effectively from mistakes. Therefore, this section begins with typical wording errors, followed by challenging words in engineering publication. You are encouraged to read writing handbooks for systematic training in vocabulary and formal writing.

7.2.1 Wordiness

Your writing may become wordy when you use more words than necessary. As a result, the writing ends up lacking effectiveness and conciseness (not clarity). There are many reasons for being wordy. Some are introduced throughout the book (*e.g.*, 4.3 and 6.5.5), others are abstract words and buzzwords. Vague modifiers and nominalization also contribute to wordiness. The following are typical word errors in engineering writing.

7.2.2 Vague words

Vague words have different meanings and interpretations. Words like *a lot*, *good*, *significant*, and *thing* are often vague words. (*See also* 4.2 and 6.5.5)

Example 7-1. Word errors (vague words)

Vague: The accuracy of the model is improved significantly.
Clear: The accuracy of the model is improved by 50%.

Comments:

The word *significantly* conveys little information. It is important to write down the number and leave it to the readers to decide its *significance.*

7.2.2.1 Nominalization errors

Nominalization is a typical mistake that non-native English writers make. It is a noun combined with a vague verb like *do, give, make*, or *perform*. For example, *perform a study* is a nominalization of study.

Nominalizations cause awkwardness in reading, although the idea remains clear. Use the specific verbs for the ideas in your mind instead of creating nominalizations using the general words.

Example 7-2. Wordiness due to nominalization

Wordy: Chapter 2 gives an introduction of the methods used in this study.
Concise: Chapter 2 introduces the methods used in this study.

Wordy: We performed a study on the effects of air pollution on respiratory diseases.
Concise: We studied the effects of air pollution on respiratory diseases.

Wordy: Avoid using emotional words in formal writing.
Concise: Avoid ~~using~~ emotional words in formal writing.
Clear: Avoid using emotional words as modifiers in formal writing.

7.2.2.2 Shift of function

Shift of function occurs when the function of a word shifts from one to another, depending on the context. Shift of function occurs frequently easily, and it may be a challenge to non-native English speakers.

Example 7-3. Word challenges (functional shift)

Talk the talk, walk the walk. *[The first talk and the first walk are verbs; the second, nouns.]*

After we present the results, we carry out in-depth analyses. After the analyses, we look after the conclusions. *[The word after shifts its function from conjunction to preposition.]*

7.2.2.3 Slangs and jargons

Jargons refer to terminology used by people in a specialized field. For example, *debug code* is a jargon in computer science, meaning test for errors or determine the causes of errors. Jargons occasionally help with conciseness but they often reduce clarity.

Slangs and jargons often lead to functional shifts of words and confusion to readers. Slangs refer to a type of informal language used only in certain situations. Like an idiom (*see* 7.4.8), a slang may be a familiar word, but it has a special meaning that may be different from its literal meaning. For instance, *chill* can mean *relax* or *calm down*; *nerd* means a socially awkward person.

Do not use jargons or slangs in academic writing, especially in publications aimed at international readers. The reason is that jargons are understood only by a unique professional group. Avoid slangs in formal writing.

Note:

Use the words and phrases that you know their precise meanings.

7.2.2.4 Localisms

Avoid words and phrases that are used only locally. Such words are usually considered informal, and readers out of the regions are unfamiliar with their meanings because of the loss of context. For example, *sub* may mean *a long sandwich* in the USA (United States of America). It may also mean *submarine* or *subscription*, depending the context. As a verb, it is shortened from *substitute.*

Example 7-4. Word errors (localisms)

"Want a sub?" "Yes, please!"

Tom subbed for Scott as goalkeeper in this soccer game.

7.2.3 Using the word "fact"

You may see many expressions containing the word *fact* in English documents, but it doesn't mean the usage of *as a matter of fact*, *in fact*, or similar phrases is correct. These phrases are wordy and can be replaced with concise items. This is especially essential to the discussion in technical writing, because you cannot force your readers to accept your judgement or opinion as a fact. Even the observation and measurement may not be the facts because there are probable errors introduced by human factors or the devices.

Example 7-5. Expressions containing the word *fact*

Incorrect: As a matter of fact, our research shows that solvent viscosity is a dominating factor for nanofiber fabrication by electrospinning.

Revised: Our research shows that solvent viscosity is a dominating factor for nanofiber fabrication by electrospinning.

As far as non-native English writers are concerned, the biggest challenge lies in the words with similar meanings or functions, but not precisely the same. There are such words in any language; these words are *translators' false friends* in linguistics. They can cause the most challenges to non-native speakers.

The next section (7.3) contains a list of challenging words (in English of course) for your usage. It is an incomplete list of words that are challenging to non-native English speakers whom I know. You are encouraged to build your own list by addition to or removal from the entries.

7.3 Challenging Words

a/an/the

The grammatical rules of articles are complicated for many non-native English speakers. Typical mistakes include omission when they are required, wordiness when they are not needed, and wrong location where they do not belong.

Example 7-6. a/an/the

Incorrect: A computation fluid dynamics (CFD) model for air flow around cylindrical object and investigation of the air velocity profile around cylinder is presented here. Turbulence is visualized to observe the effects of flows.

Correct : Presented here is a computation fluid dynamics (CFD) model for *the* air flow around *a* cylindrical object and *a* visualization of ~~the~~ air velocity around the cylinder. Turbulence is visualized to observe the effects of flows.

a lot/many

Do not use *a lot* in formal writing; instead, use *many* or *numerous* for estimation. It is even better to give specific values when they are available.

above/aforesaid/aforementioned

Above, aforesaid, and *aforementioned* are vague words. Avoid using these words to refer to the preceding nouns and pronouns. Simply repeat the nouns or pronouns for clarity.

absolutely/definitely/entirely/completely/unquestionably

Avoid these words because they are redundant intensifiers when they are used for the meaning of *very* or *much.*

Example 7-7. Redundant intensifiers (*definitely*)

We are ~~definitely~~ certain that it was the catalyst fouling which caused the damage.

Clean air is ~~absolutely~~ necessary for a healthy and productive economy.

accuracy/precision

Accuracy means being correct without a single error. *Precision* is refinement in a measurement, calculation, or specification. For instance, 3.14159 is more precise than 3.14. *Precision* does not necessarily mean *accuracy.* *See* engineering statics books for related basics.

activate/actuate

Activate something mean *make something active. Actuate* is usually limited to mechanical devices.

adapt/adopt

Adapt means *to adjust to new conditions* or *to modify for a new purpose. Adopt* means *to accept* or *to take on.*

Example 7-8. adapt/adopt

We must adopt the latest safety guideline to adapt to the new working environment.

affect/effect

Affect is a verb, and *effect* can function as both a noun and a verb. When *effect* acts as a noun, it means *result*. When *effect* is used as a verb it means *result in*, *cause*, *make*, *produce* and the like. Non-native English writers may want to avoid using *effect* as a verb, if possible, in technical writing.

all together/altogether

Altogether means *completely*, *everything considered*, or *on the whole*. *All together* means *all in one place*, *all in a group*, or *all at once*.

also/too/in addition

Also, too, and *in addition* have very close meanings. However, do not use *also* to open a sentence or to connect two elements in a sentence; instead use *in addition* (which is a preposition). *Too* ends a sentence.

Example 7-9. also/too/in addition

Incorrect: Heat exchangers are used in power plants, also they are used in other systems.

Incorrect: Heat exchangers are used in power plants. Also, they are used in other systems.

Correct: Heat exchangers are used in power plants, and they are used in other systems.

Correct: Heat exchangers are used in power plants; they are used in many other systems too.

Correct: Heat exchangers are used in power plants. In addition, they are used in other systems.

amount /number/quantity

Amount is used for mass nouns that cannot be counted by number (cement, electricity, paper, water). *Number* is used with countable nouns (device, book). *Quantity* can be used with both mass nouns and countable nouns. *Quantity* is more formal than *amount* or *number*.

Example 7-10. amount /number/quantity

The amount of cement is determined before it is mixed with sand.
The number of bags of cement is determined before mixing the cement with sand.
The quantities of cement and sand are determined before mixing them.

and/or [used together]

Avoid *and/or* in formal writing; instead use *or … or… or both.*

Example 7-11. and/or

Confusing: You can use WebEx and/or Zoom for video conferences.
Revised: You can use WebEx *or* Zoom *or both* for video conferences.

as/because/since

As, because, and *since* are commonly used to introduce subordinate clauses that express causes. *Because*, more commonly used than *as* and *since*, introduces the most specific reason. *Since* is a weak substitute for *because*, but stronger than *as.* However, *since* can be used for additional factors emphasizing on circumstance or time. Avoid using *as* to indicate cause in formal writing, although it has this function informally.

Example 7-12. as/because/since

Since the paper has been published, we must prepare a list of errata for the readers because there are several typos in the text. *[Since indicates a change of circumstance; because indicates the cause of action.]*

as such

Avoid the phrase *as such* in formal writing. Instead, use *thus, therefore, etc.*

as well as / and

Either *as well as* or *and* means *in addition*, but they are used with different verbs in sentences.

Example 7-13. as well as / and

Both temperature and pressure of an ideal gas can affect its volume.
Temperature as well as pressure of an ideal gas can affect its volume.

average/mean

Both *average* and *mean* are statistical terms; they are usually confused with one another, especially to non-native English writers. There are several types of *means* in statistics, and *average* (or arithmetic *mean)* is the simplest form of *mean.* Make sure you know precisely what kind of *means* you are referring to in your writing. (See *significant* as a modifier.)

begin/start

Begin is an irregular verb, and *start* can act as a noun and a regular verb. *Begin* is more formal than start (as a verb), but they have the same meaning. *Start* and *beginning* mean the same thing when they are used as nouns.

beside/besides

Beside means *in a position immediately to one side of*; *besides* means *as well* or *in addition to*.

Example 7-14. beside/ besides

Besides the agitated heater, the oven and the water bath are located *beside* each other on the bench.

between/among/amongst

Between is normally used when the number involved is two, whereas *among* and *amongst* are for three or more items. *Amongst* and *among* have the same meaning, but *among* is more common in American English. *Amongst* is a British and Canadian spelling of *among*.

can/may

Can refers to capability; *May* refers to possibility.

Example 7-15. can/may

We certainly can finish the test, but we may not meet the deadline

cannot/can't/can not

Cannot is one word. Do not use *can not* except for *can not only*. *Can't* is a contraction of *cannot*, and it is meant for informal communication only.

Example 7-16. cannot/can't/can not

Wrong: Carbon dioxide can not reach its supercritical state if the temperature is below 31.1 °C.

Informal: Carbon dioxide can't reach its supercritical state if the temperature is below 31.1 °C.

Correct: Carbon dioxide cannot reach its supercritical state if the temperature is below 31.1 °C.

Wrong: Your article cannot only report the results; you need to add the discussion part.

Correct: Your article can not only report the results; you need to add the discussion part.

compare/contrast

You *compare* to establish similarities or differences or both, and you *contrast* to point out only the differences (not the similarities). When they function as verbs, *compare* is followed by *to*, and *contrast* is followed by *with*. When *contrast* is used as a noun, it is followed by *between*.

Example 7-17. compare/contrast

We compared our results to earlier ones.

Our results contrast sharply with earlier ones. *[Contrast is a verb]*

There is a sharp contrast between our results and the earlier ones. *[Contrast is a noun]*

complement/compliment

Complement means a thing that completes or brings to perfection. It acts as either a noun or a verb. *Compliment* means a polite expression of praise. *Compliment* can be used as a noun or verb.

Example 7-18. complement/compliment

Our results complement the earlier findings reported earlier in literature.

All reviewers compliment the high quality of the work and the manuscript. *[compliment as a verb]*

Thank you! I consider that as a compliment. *[compliment as a noun]*.

complete/finish

Complete and *finish* are confusing words, especially to non-native English speakers. *Complete* is a positive word, which means that something has no missing parts. *Finish* means *to end* or *stop*. In a study that involves chemical reactions, for example, that the reactions *finish* does not mean the reactions are *complete* (all reactants are consumed.)

complex/complicated

A *complex* system has many components. Complexity indicates the level of components in the system but does not necessarily evoke difficulty. *Complicated* is used for something causing a high level of difficulty.

continual/continuous

Continual means something happens over and over. It repeats many times. *Continuous* implies that something continues without interruption.

Example 7-19. continual/continuous

Effective technical writing requires continual learning and practice.

The fluids are usually assumed continuous, rather than discrete, in computational fluid dynamics models.

criteria / criterions

Both *criteria* and *criterions* are plural nouns; their singular form is *criterion*, which means *a standard by which something may be judged*. However, the word *criteria* is more commonly used in technical writing.

data/datum

In formal writing, *data* is a plural noun, and its singular form is *datum*. In informal writing, *data* can be used as a singular noun.

Example 7-20. data/datum

Informal: The experimental data in Fig. 6 supports the argument.

Formal: The experimental data in Fig. 6 support the argument.

definite/definitive

Definite means known for certain; *definitive* means conclusive. Both imply something precisely defined.

Example 7-21. definite/definitive

There is a definite correlation between global temperature and the carbon dioxide concentration in the atmosphere; however, it might not be definitive to say that global warming is solely caused by CO_2.

despite/in spite of/although/even though

In spite of and *despite* have a similar meaning to *although* or *even though*. Both *despite* and *in spite of* introduce nouns; *although* and *though* introduce subordinate clauses. *Despite* is a little more formal than *in spite of*; they are used to emphasize efforts to avoid blames.

Example 7-22. despite/in spite of/although

In spite of /Despite our great efforts, the pilot tests failed.
The pilot tests failed, although/even though we tried our best.

different from/different than

Different from typically precedes a *noun* or noun form, while *different than* may be followed by a *clause*, to complete an expression.

Example 7-23. different from/different than

Our approach to the problem is different from those reported earlier.
We took an approach that is different than what others did earlier.

due to/because of

Due to and *because of* both mean *caused by*. However, *due to* modifies the nouns, and *because of* should modify the verbs.

Example 7-24. due to/because of

Wrong: Your promotion was *because of* your outstanding performance.
Correct: Your promotion was due to your outstanding performance.
Wrong: You were promoted *due to* your outstanding performance.
Correct: You were promoted because of your excellent performance.

economic/economical

Economic refers to how money works; *economical* means *not wasteful*.

Example 7-25. economic/economical

The technology was developed in an economical way, but an economic analysis is necessary before technology transfer.

emphasize/emphasis on

Emphasize is a verb, and *emphasis* is a noun. *Emphasize* something means *put emphasis on* something. Do not use emphasize on. (*See* 6.4)

everybody/everyone/every one

Everybody and *everyone* are interchangeable, and they are used with persons. They both appear to be singular, but meanings can be plural. *Every* and *one* emphasizes each individual.

Example 7-26. everyone/every one

Everyone of the co-authors must have some contributions to the work *[actually meaning all]*.

Every one of the co-authors contributed to the work *[emphasize individual]*.

few/a few/fewer; little/a little/less

Few, *a few* and *fewer* are used with countable nouns; *little*, *a little* and *less* are used with mass nouns. *Few* or *little* sounds less (even negative) than *a few* or *a little* does.

first/firstly; second/secondly; third/thirdly ...

Avoid *firstly* and use *first* for formal communications. The same is true with other ordinal numbers.

flammable / inflammable / non-flammable

Both *flammable* and *inflammable* mean *able to burn* even though they look like opposites. Use *flammable* instead of *inflammable* to avoid confusion to international readers. *Non-flammable* is the opposite of *flammable.*

imply/infer

Imply means *suggest* or *hint [implicitly]*; *infer* means *draw a conclusion* from evidence *[explicitly]*.

in/into; in/within

In means *inside of*, normally an area or space. *Into* is connected to the movement of something. *Into* typically follows a verb like *go, come, run, etc.* *Within* stresses that something is not further than a prescribed area or space or not later than a time, but *in* does not indicate emphasis.

Example 7-27. in/into/within

The sensor is in the device.

The sensor is put into the device.

The test will be finished in 20 minutes. *[Estimate the time needed]*

The tests must be finished within 20 minutes. *[Set the time limit]*

in order to/to

Use *to* instead of *in order to*, although they have the same meaning. You occasionally can use *in order to* for adjusting the pace of presentation or balancing the sentence length.

inside/inside of

The word *of* is *redundant* in the phrase *inside of*. Use *inside* only in formal writing.

in this paper/study/work/research

The *study* is the *research* or *work* or the like that you did. The *paper* is one of many ways that you present your *work* (or the *study*) to others; the *paper* is what you are writing. Do not overuse the word *in this paper* or *in this study*. Either phrase should not appear more than three times in an article.

keyword/key word

Keyword is a search-related word either online or in an index. *Key word* means *important word.*

lay/lie

You can write, *lay something* and *something lies*; *lie* and *lay* are not interchangeable. Their past-tense forms are even more confusing. The past-tense forms of *lay* and *lie* are *laid* and *lay*, respectively. (*Lied* is the past-tense form of *lie* when it means *not telling the truth.*)

Example 7-28. lay/lie

We normally lay the filter samples on the bench and the samples lie in the lab for 24 hours before analysis.

We laid the filter samples on the bench; they lay in the lab for 24 hours before analysis.

like/as

Both *like* and *as* are used in comparisons. *Like* is a preposition and it introduces a noun that is not followed by a verb; while *as* is a conjunction followed by a verb.

maybe/may be

Maybe means *perhaps* or *possibly* (as adverb) or a mere possibility or probability (as noun). *May be* is a phrase that means *something might.*

on/onto/upon

The relationship between *on* and *onto* is similar to that between *in* and *into*. *On* stresses a position of rest; *onto* implies movement to a position. *Upon* emphasizes movement or a condition.

Example 7-29. on/onto/upon

The revision of your manuscript will start upon completion of the additional tests.

only/solely

Only and solely are often interchangeable in technical writing, but *solely* emphasize exclusivity. The word *only* or *solely* should precede the word or phrase it is intended to modify; otherwise, the meaning of the sentence may change.

Example 7-30. only/solely

Only we reported the effects of temperature on alkane biofuel *conversion rate. [Nobody else reported]*

We only reported the effects of temperature on alkane biofuel conversion rate. *[We did not report other information.]*

percentage/percent/percentile

Percentage is a number that is written out of 100, but it cannot be used with a number. *Percent* is normally used to quantify *percentage*. A *percentile* is a *percentage* of values found below a specific value.

Example 7-31. Percentage/percent/percentile

Two out of 20 students received grades of 150 or lower out of a total of 200 in a test. The percentage of students who received grades of 150 or lower grades is 10 percent (10% calculated from 2/20); the 90th percentile on the test is 150.

precision/preciseness

Precision shows the ability to reproduce data consistently or the number of digits indicating the reliability of a measurement. *Preciseness* shows the quality of being accurate, especially about details. *Precision* and *preciseness* are synonyms, but *precision* is more frequently used for measurements than *preciseness* is.

Example 7-32. precisions/preciseness

You can improve the precision of measurement by changing the meter range from 1-1000 nm to 10-100 nm.

You can further improve the quality of your writing by checking the preciseness of punctuation.

principal/principle

Principal can be a noun or an adjective, meaning *main, primary, most important,* or *influential*, or a person of such importance. *Principle* is a noun only and it means *a rule or code of conduct* or *the basics of truth.*

Example 7-33. Principal/principle

The principal objective of this study is to understand the principle of air conditioning.

raise/rise/arise

Both *raise* and *rise* mean "move up." *Raise* is a regular transitive verb, followed by a noun (object); *rise* is an irregular intransitive verb that follows a noun or the like. You can write *raise something* and *something rises. Arise* mean *to begin to occur* or *to stand up.*

Example 7-34. raise/rise/arise

When you raise your arm, its gravitational potential energy rises. When you arise from your chair, your gravitational potential energy rises too.

really/actually

Avoid using *really* or *actually* as intensifiers for emphasis in formal writing. *[See overused intensifiers.]*

reason is/ because

Never use *the reason is because*, which is redundant. You can use *because* or *the reason is that.*

regarding/with regard to/in regard to

Regarding, in regard to, and *with regard to* all mean *with respect to* or *concerning.* There are interchangeable. However, there are no such phrases as ~~*in regards to*~~ or ~~*with regards to*~~.

Regardless of /irregardless

Regardless of means *without regard or consideration for.* It is a negative and disqualifies *irregardless* as a standard word, because *irregardless* expresses a double-negative. Use *regardless* instead of *irregardless* in your writing.

Example 7-35. regardless of

This paper is technically valuable to the readers in the field regardless *of* its poor writing.

respective/respectively

Respective (adjective) means *belonging* or *relating separately to each of two or more individuals. Respectively* (adverb) means *in the order mentioned. Respectively* often appears at the end of the sentence.

Example 7-36. respective/respectively

Wrong: Eqs. 1-3 can be respectively used for the calculation of A, B, and C.

Wrong: Eqs. 1-3 can be, respectively, used for the calculation of A, B, and C.

Right: Eqs. 1-3 can be used for the calculation of, A, B and C, respectively.

Wrong: The temperatures at the inner surface and outer surface of the tube are respectively 550 °C and 560 °C.

Wrong: The temperatures at the inner surface and outer surface of the tube are, respectively, 550 °C and 560 °C.

Right: The temperatures at the inner surface and outer surface of the tube are 550 °C and 560 °C, respectively.

shall /will

You can just use *will* and ignore *shall* for future tense in technical writing. Shall has a very strong tone, and it is occasionally used in legal documents.

some time/sometime/sometimes

Avoid *some time*, *sometime*, and *sometimes* in technical writing, where precision and accuracy are expected by readers in engineering. *Some time* means a period of time; *sometime* means a time to be determined. *Sometimes* mean *occasionally* at unspecified times. They all can be used in verbal communication or informal writing.

Example 7-37. sometimes/sometime/some time

Vague: We waited for some time to ensure that chemical reactions stopped.
Vague: Let's get together sometime so we can analyze the data Tom collected.
Vague: The lab director sometimes comes to the lab to check the safety.

such as/for example/ *etc.* /and so on/and so forth/and the like

Such as, for example, etc., and so on, and so forth, and *and the like* all can be used with incomplete lists, but none of them should be used with a complete list.

Such as and *for example* can introduce an incomplete list; *etc., and so on, and so forth,* and *and the like* are used at the end of an incomplete list. However, using any two of these phrases together causes redundancy. *Such as* and *etc.* are often misused by non-native English writers.

use/utilize/usage

Both *use* and *utilize* are verbs. Do not use *utilize* in technical writing, because it appears to be pretentious and vague. *Usage* is a noun; it means the action of *using something* or the fact of *being used.*

whether/if

Use *if* to introduce a conditional sentence and *whether* to communicate the notion of choice. In the phrase *whether or not*, the words *or not* are redundant. The phrase *as to whether* is redundant too.

Example 7-38. whether/if

The solvent selectivity depends on whether ~~or not~~ it contains salts.

The solvent selectivity is greater if the salt concentration is higher.

while/although/whereas

Using *while* often results in ambiguity when it is substituted for connectives like *and*, *but*, *although*, and *whereas*. Avoid using *while* for this purpose. Instead, use the words that you hope while to substitute. Use *while* only when it means "*during the time that.*"

Example 7-39. while/although/whereas

I wrote this book while the university was closed because of the COVID-19 Pandemic.

whose/of which

Whose is used with persons; *of which* is used with inanimate objects. *Whose* is occasionally used with an inanimate object to achieve conciseness.

Example 7-40. whose/of which

This book, the title of which is plain, was written by Dr. Tan, whose employer is the University of Waterloo.

7.4 Phrases

A phrase is a group of words. Meaningful phrases act as adjectives, adverbs, nouns, or verbs in clauses and sentences. Unlike clauses, phrases cannot state an idea because it contains neither a subject nor a predicate. Phrases provide context within a clause or sentence where the phrases are used. (*See also* 6.1).

Example 7-41. Adjective phrase and adverb phrase

Adjective phrase:

This item has been received already *[Without phrase]*

This item on the list has been received already. *[With phrase]*

Adverb phrase:

We have received this item on the list in a perfect shape.

7.4.1 Noun phrases

A noun phrase functions as a subject, object, or prepositional object in a sentence. It consists of a noun and its modifiers.

Example 7-42. Noun phrases

Before the publication of your articles, you may have to sign the agreement to release your copyrights.

7.4.2 Gerund phrases

A gerund phrase is originated from a verb ending with *-ing*. It is different from a noun phrase. The gerund phrase can act as either a subject or an object in a sentence.

Example 7-43. Noun phrases

Subject:

Writing this book is driven by my enthusiasm in higher education.

Writing a journal article requires certain skills and good planning.

Object:

All engineering PhD students enjoyed writing journal articles.

7.4.3 Verb phrases

A *verb phrase* is formed by an *auxiliary verb* followed by the *main verb*. Some words may appear between the auxiliary verb and the main verb. In the following sentence, for example, *am* is the auxiliary verb, and *writing* is the main verb.

Example 7-44. Verb phrases

I was writing this book when the school was closed because of COVID-19 pandemic.

This item on the list should have been received. *[In reality it has not.]*

7.4.4 Verbal phrases

Verbal phrases are formed with the phrases and verbals. A verbal phrase can act as a modifier, an object, or a complement. A verbal is a word that combines characteristics of a verb with those of an adjective or a noun. Non-native English writers are often confused with the form of a verbal. The best approach to this challenge is reading English materials, listening to native speakers, and practising what you learn. You are also encouraged to consult reference books for non-native English speakers.

7.4.5 Prepositional phrases

Prepositional phrases normally modify nouns or verbs or the like in the same sentence. They may act as adjectives following the nouns that they modify.

Example 7-45. Prepositional phrases

The power will turn off automatically after the temperature reaches 350 °C.

Agricultural biomass waste with a high cellulose content can be converted into biofuel by hydrothermal conversion.

Overuse of prepositional phrases can lead to wordiness or ambiguity. For clarity, you can use modifiers, which are more economical, for the same purpose as prepositional phrases. *See* Example 7-46.

Example 7-46. Prepositional phrases and modifiers

Wordy: The data presented with solid black circles on the dashed lines in Figure 3 are obtained experimentally in the lab. *[prepositional phrases used]*

Concise: The data in Fig. 3, presented using solid black circles on the dashed line, are experimental results. *[Modifier used]*

Vague: The curve for model and that for experiment in red color agree with each other. [*Not clear whether one or both in red color.*]

Clear: The curve for model in red color and that for experiment agree with each other.

7.4.6 Participial phrases

For clarity, it is important to describe the relationship between a participial phrase and other elements in the same sentence. Otherwise, the phrase becomes dangling participial phrase, which modifies nothing in the sentence. (*See* 6.5.5.4)

Example 7-47. Participial phrases

Dangling: The system having the highest thermal efficiency chosen for further studies at the pilot scale.

Correct: The system having the highest thermal efficiency was chosen for further studies at the pilot scale.

Dangling: Being unsatisfied with the accuracy, the device was replaced with another one. *[Dangling phrase: The device cannot be unsatisfied; it must be person.]*

Correct: Being unsatisfied with the accuracy of the data, we replaced the device with another one for further studies at the pilot scale.

7.4.7 Infinitive phrases

An infinitive phrase begins with the word *to*, followed by a verb. The infinitive phrase indicates the purpose of an action. Note that a prepositional phrase also begins with *to* but followed by a noun or a pronoun. A prepositional phrase normally indicates a destination for an action.

Example 7-48. Infinitive and prepositional phrases

Infinitive: To meet the deadline, I am willing to work overtime every day for the coming week.

Prepositional: He went to school for education. *[school is a noun.]*

Both: He went to the library to prepare for the final exam. *[prepare is a verb.]*

7.4.8 Idioms

Among the variety of phrases, idioms may be as challenging as slangs to non-native speakers of English. Like slangs, idioms are phrases that cannot be *literally* interpreted. For example, *run a test* means *to arrange for someone or something to be tested* – nobody is running, literally. This type of expression is culture specific; therefore, non-native English speakers need to memorize them and increase their vocabulary over time.

Table 7-1 lists the idioms commonly used in engineering communication. These idioms can make writing more vigorous, creating a natural connection between the writers and the readers.

Use idioms with caution and make sure you know the exact meaning first. Furthermore, they may cause confusion, even frustration, to non-native English readers. An international reader may stop reading and check dictionaries or search online information to interpret the precise meaning of idioms.

Table 7-1. Commonly used idioms in technical communication

Idiom	Meaning
agree to	consent
agree with	in accord
Back to the drawing board	When an attempt fails and it's time to start all over
call off	cancel
cross out	draw a line through
do over	repeat a task
figure out	solve a problem
find out	discover information
get through with	finish
hand in	submit
hand out	distribute
keep on	continue
leave out	omit
look up	research a subject
Picture paints a thousand words	A visual presentation is far more descriptive than words.
run out of	deplete supply
watch out for	be careful

~~~

You have finished revising your draft following the guidelines on paragraphs, sentences, phrases, words, *etc.* The document after extensive revisions becomes clear, concise, and coherent at this point. However, you can further improve its clarity and preciseness by checking the punctuation marks, which are used to separate sentences and sentence elements. It may not sound as important as the preceding chapters, but punctuation marks do affect the overall quality of your final document. Check them with care following the guidelines in the next chapter.
~~~

8 Punctuation

8.1 Names and Marks of Punctuation

Punctuation is an important part of sentences. Understanding punctuation is essential to communication with clarity and preciseness. Table 8-1 lists the 13 punctuation marks commonly used in academic technical writing in English. They are presented in this chapter following an alphabetic order instead of the order of importance in writing.

Table 8-1. Names and marks of punctuation

Name	**Mark**	**Name**	**Mark**
Apostrophe	’	Parentheses	( )
Brackets	[ ]	Period	.
Colon	:	Question mark	?
Comma	,	Quotation marks	“ ”
Dash	--	Semicolon	;
Exclamation mark	!	Slash	/
Hyphen	-		

8.2 Apostrophe

An apostrophe can be used

- To show possession of nouns (*e.g.* the *paper’s title*; the *authors’ addresses*).
- To form plurals of lowercase letters and abbreviations (*e.g.*, *three y’s*).

  You do not need to use apostrophes with a plural on capital letters or numbers such as *As, PDFs,* and *Drs. See*
- To indicate the omission of letters *(e.g. don’t, int’l).*

Using apostrophe for omission of letters is not encouraged in formal writing, but the other two are frequency used in engineering writing.

Example 8-1. Apostrophes

You don't need to write down all authors' contact information. *[Avoid omission of letters in formal writing.]*

He wrote three *i's* in the same equations with different meanings.

Replace all *e.g.*'s with *for example*'s would not help achieve conciseness.

The best student in class received four *As* and two *Bs* in his courses.

This book was written in early 1990s by two PDFs. (*See* 4.14)

8.3 Colons

Colons (:) are used in sentences, numbers, times, *etc.* for a verity of functions. The colon acts as a pause and may catch the readers' attention prior to introducing what follows. Colons can also be used with numbers indicating times or numerical ratios.

Example 8-2. Colons

An engineering research project should meet two criteria: the scope of research is new to the field and the contributions are important to the society.

A graduate thesis should contain at least five parts: literature review, objectives, methodology, results and discussion, and conclusions.

Figure 5 shows the strong correlation between traffic and low air quality for between 7:30 a.m. and 9:30 a.m.

Air is a mixture of nitrogen, oxygen and other gases at a volumetric ratio of 78:21:1

Don'ts for colons

- Don't capitalize the word immediately following colons.
- Don't insert a colon between a verb (or a preposition) and its objects. A non-native English writer might think it emphasizes what follows; instead, it becomes a grammatical error.
- Don't immediately follow *including, such as, for example, etc.* with a colon.

Example 8-3. Don'ts for colons

Wrong: Common types of catalysts include: enzymes, acid-base catalysts, and surface catalysts.

Right: Common types of catalysts include enzymes, acid-base catalysts, and surface catalysts.

Wrong: This manuscript may be submitted to: *Science, Nature, or Nature Nanotechnology.*

Right: This manuscript may be submitted to *Science, Nature, or Nature Nanotechnology.*

Wrong: This article can be submitted to many prestigious journals such as: *Science, Nature, and Nature Nanotechnology.*

Right: This article can be submitted to many prestigious journals such as *Science, Nature, and Nature Nanotechnology.*

8.4 Comma

Commas and *periods* are the most frequently used punctuation marks. A period ends a sentence, and a comma indicates a brief pause before the sentence ends. Commas have the following functions in sentences: linking independent clauses, introducing elements, enclosing elements, quotations, separating items, clarifying and contrasting meanings, indicating omissions, *etc.*

8.4.1 Replacing verbs

Experienced writers may replace the verb in a sentence with parallel structure. You may have noticed that I occasionally did so in this book. (*e.g., The passive voice is used frequently in some languages, but in others, not at all.*) It is grammatically correct, but it chops the sentence into pieces. The omission of verbs may cause confusion to non-native readers of English. Thus, do not replace verbs in a sentence with commas, especially in technical writing.

Example 8-4. Comma replacing verbs

Correct: Some researchers are prolific in publication; others, not at all.

Clear: Some researchers are prolific in publication; others are not at all.

8.4.2 Linking independent clauses:

A comma mark is needed immediately before a coordinating conjunction that links independent clauses. *See* Table 6-1 for more examples

Example 8-5. Comma linking independent clauses

Your description of procedure is clear, but you omit the key specifications of the devices.

8.4.3 Enclosing elements

Commas can be used to enclose nonessential elements or appositive phrases in a sentence. If the elements enclosed by the commas were removed from the sentence, the rest of the elements still construct a valid sentence, and the primary idea of the sentence survives the removal.

Example 8-6. Comma enclosing elements

Non-restrictive clause:

From the results in this paper, which were presented in Section 4, they drew the following conclusions.

The research team, working day and night, finished the project on time.

An appositive phrase:

The only book that Margaret Mitchell authored, *Gone with the Wind*, was immensely popular when first released.

Interrupting words or phrases:

The revised manuscript, therefore, is accepted for publication.

The revised manuscript, however, cannot be accepted for publication.

Comments:

The words enclosed with commas (*therefore, however, for example*) slow down the pace of presentations, and they interrupt the continuity of thought.

8.4.4 Introducing elements

A comma is needed after an introductory word, phrase, or clause to set a brief pause before the main part of the sentence begins. This is especially important to pace control for the sentences with long introducing elements. *See* Appendix for the list of transitional words and phrases.

Example 8-7. Comma introducing elements

Transitional word:

However, the SO_2 removal efficiencies met our expectations.

Transitional phrase:

For example, the SO_2 removal efficiency reached 99.5% when the temperature was 40 °C.

Introductory long phrase:

During the final step of the pilot tests, the cobalt based solvent did not show the NOx removal efficiencies as expected.

Pilot tests finished, and we realized that the cobalt based solvent did not show the NOx removal efficiencies as expected.

Introductory clause:

Because the hospital did not have enough ventilators for all patient during the of COVID-19 pandemic, health professionals had to make difficult decisions on who got the life-saving machines.

Note:

A sentence that begins with an introductory element may change the emphasis of the sentence. (*See also* 6.4.1) In addition, the comma is optional if it follows a short phrase (Example 8-8). However, comma should follow an absolute phase.

Example 8-8. Comma omission from short phrases

In pilot test the cobalt based solvent did not show the NOx removal efficiencies as expected. *[short phrase]*

8.4.5 Separating parallel elements

Commas are used to separate parallel elements (words, phrase, and clauses) in a sentence. For this application, comma before the last element can be omitted occasionally. However, including the last comma improve clarity, especially when the last comma comes right before the word *and* or *or*. (*See also* 4.4, 5.1, and 6.3.1)

Example 8-9. Comma for separating elements in parallel series

Word series:

Vague: Apple, Blackberry, Huawei and Samsung were the leading smartphone producers in the world. *[A comma following Huawei is needed for clarity; otherwise, Huawei and Samsung might appear to be one company.]*

Clear: Apple, Blackberry, Huawei, and Samsung were the leading smartphone producers in the world.

Phrase series:

To complete this course, students must learn basics in the classroom, observe their applications on a training site, give presentations to the class, and submit final reports by email.

Clause series:

A modern university education is all about students: students are motivated in learning, staff members are keen to serving the students, and faculty members are passionate in teaching.

8.4.6 Using commas with numbers

Commas are commonly used with addresses, dates, and numbers in technical writing. *See also 9.7, Formatting Numbers.* For example, the Arabic number of *one million, two hundred thousand, and three hundred* is 1,200,300. However, many countries use periods (1.200.300) or single spaces (1 200 300) for the same number – they are rarely use in English documents.

Example 8-10. Comma used with numbers, dates and addresses

The 1,234-page long document was released to the public on April 1, 2020 in Waterloo, Ontario, Canada.

8.4.7 Don'ts for commas

- Don't insert a comma wherever you feel like a brief pause. Use of commas without challenging the grammatical rules of sentence construction in scholarly technical writing. (*See* 6.1)
- Don't insert a comma between a subject and its verb or between a verb and its object.

  A non-native English writer might think it emphasizes what follows; instead, it becomes a grammatical error. Additionally, this "comma intrusion" interrupts the smooth flow of thoughts.
- Don't insert comma in a compound element consisting of only two elements.

Example 8-11. Comma challenges (1)

Wrong: The director of the laboratory, and the administrative staff are responsible for the safety of the researchers working in the laboratory.

Right: The director of the laboratory and the administrative staff are responsible for the safety of the researchers working in the laboratory.

Wrong: The principle investigator designed the test apparatus, and gave a demonstration to the team members.

Right: The principle investigator designed the test apparatus and demonstrated it to the team members.

- Don't enclose a coordinating conjunction (such as *and, but, or*) with commas.

Example 8-12. Comma challenges (2)

Wrong: The principle investigator designed the test apparatus, and, gave a demonstration to the team members.

Right: The principle investigator designed the test apparatus and demonstrated it to the team members. *[It is clear who gave a demonstration to the team members.]*

Wrong: Do not insert a comma between an object and its verb, or, a verb and its object.

Right: Do not insert a comma between an object and its verb or a verb and its object.

- Don't omit commas.

  Commas should immediately follow conjunctive adverbs or conjunctive phrases that join independent clauses. (*See also* Table 6-1. Conjunctions)

Example 8-13. Comma challenges (3)

Wrong: The authors presented a timely work; nevertheless it does not match the scope of the journal.

Right: The authors presented a timely work; nevertheless, it does not match the scope of the journal.

- Don't join two independent clauses with one single comma.

  Many non-native English writers run two independent clauses together by using a comma instead of a period. This results in a *run-on sentence* or a *comma splice*.

Example 8-14. Comma challenges (4)

Incorrect: We turned on the power, the heater started heating the reactor.

Correct: We turned on the power. *The* heater started heating the reactor.

Correct: After we turned on the power, the heater started heating the reactor.

Correct: We turned on the power, and the heater started heating the reactor.

Correct: We turned on the power; the heater started heating the reactor.

8.5 Dashes

A dash is indicated by a short line longer than a hyphen or by two consecutive hyphens. It can stress on linkage, separation, emphasis, informality, or abruptness. However, it is used occasionally in technical writing. You can write an engineering document without dash marks.

8.6 Ellipsis

The ellipsis mark (. . .) indicates the omission of contents. It is rarely used in scholarly technical writing because omission may reduce clarity. You should avoid quotation in technical writing. In this book, however, I use several ellipsis marks in Table 5-2 to show the omission of table contents, which are not essential to my writing in this book.

8.7 Hyphen

The hyphen (-) is used primarily in compound words; it can also act as a linkage or an emphasis to improve the clarity of writing. You can find numerous hyphenated words in this book, but the hyphen has more functions, which are introduced as follows.

8.7.1 Hyphenated words

Two or more words joined by one or more hyphens form a compound word. For example, *state-of-the-art*, *three-quarters*, and *forty-eight.* Some hyphenated words are standard and collected in dictionaries. You can create new hyphenated words as needed when you write.

8.7.1.1 Multi-word modifiers

You can create two- or three-word "adjective" modifier, which has one single meaning. Multi-word modifiers precede nouns.

Example 8-15. Hyphenated words as modifiers

Springer Verlag is a well-thought-of publisher for engineering books and periodicals.

8.7.1.2 Series

You can used hyphens in a series of unit modifiers that precede the same noun. By doing so, your writing flow smoothly with brevity.

Example 8-16. Hyphenated words in series

You can create two- or three-word "adjective" modifiers, which have single meanings. *[I used this sentence right above Example 8-15.]*

8.7.1.3 Prefixes

A hyphen can be used between a prefix and a proper noun or a word acting as a noun (*e.g., pre-mixing*). It is also used when the prefix ends in the same vowel that the root words starts with (*semi-industrious, re-enter*).

Example 8-17. Hyphenated words as prefixes

Power plants reduce air emissions by pre-, in-, and post-combustion control technologies.

8.7.2 Hyphens for clarity

Use hyphen with caution to improve clarity and avoid ambiguity. The hyphens inserted into different words may lead to different modifiers.

Example 8-18. Hyphenated words for clarity

Ambiguous: The City of Waterloo is investigating the feasibility of a biochemical waste management facility.

Revised: The City of Waterloo is investigating the feasibility of a biochemical waste-management facility.

Revised: The City of Waterloo is investigating the feasibility of a biochemical-waste management facility.

8.8 Parentheses

Words, phrases, or sentences enclosed by parentheses normally clarify or define the preceding texts. The materials in parentheses can be removed from the sentence and the idea of the sentence should survive the removal. The information contained in the parentheses is non-essential to the sentence, but it may be helpful to some readers.

Example 8-19. Parentheses in sentences

Pre-combustion (prior to combustion) air cleaning is more economical than post-combustion (after combustion) air cleaning.

Don'ts for parentheses:

- Don't use parentheses with numeral for a number that is described with the words. This is considered redundant and interrupt the flow of reading.

Example 8-20. Parentheses errors

The abstract of a Master's thesis should be less than two ~~(2)~~ pages.

- Don't use parentheses within parentheses; use brackets instead.

Example 8-21. Parentheses errors

Incorrect: (Source: A manuscript submitted to Atmospheric Environment (January 2013))

Incorrect: [Source: A manuscript submitted to Atmospheric Environment (January 2013)]

Correct: (Source: A manuscript submitted to Atmospheric Environment [January 2013])

8.9 Period

Again, periods and commas are the most frequently used punctuation marks in writing. A period ends a full sentence. In addition, periods are used in decimal points with numbers (*e.g.*, 31.5 degrees Celsius) and with lowercase abbreviations (*etc.* and a.m.)

Do not end a sentence with more than one period mark. This is often confusing to non-native English writers, especially when the sentence ends with an abbreviation that ends with a period. In addition, avoid a period mark where it does not belong; otherwise, it creates sentence fragments. *See also* 6.5.1.

Example 8-22. One period mark only at the end of a sentence

The group meeting will start at 8:00 a.m.

Scholarly publications include scientific journal articles, non-fiction books, theses, *etc.*

8.10 Question Marks

The question mark is rarely used in engineering publications, which are mostly narrative. However, it may appear in direct quotations or it is occasionally used to provoke discussion.

8.11 Quotation Marks

You can use quotation marks to enclose words, phrases, clauses, sentences, and paragraphs. When words, phrases and clauses are enclosed, the quotation marks indicate a meaning of "*so-called*". When sentences and paragraphs are enclosed, however, they indicate direct quotations.

Example 8-23. Quotation marks meaning "so-called"

Many people believe that hydrogen is a "zero-emission" fuel.

8.11.1 Titles and quotation marks

The actual titles of the documents are not enclosed by quotation marks. In the body text, however, quotation marks are needed to enclose titles of most publications, except for titles of books and periodicals.

Example 8-24. Quotation marks enclosing titles

Einstein's doctoral thesis, "A New Determination of Molecular Dimensions", published in 1905 was only twelve-page long.

"Basics of Gas Combustion" is the third chapter of the book titled *Air Pollution and Greenhous Gases.*

Note:

Do not enclose the titles of books and periodicals with quotation marks; these titles should appear in *italic* font.

8.11.2 Punctuation in quotation marks

Materials directly quoted remain in the quotation marks, except for the following two fixed rules:

- Colons and semicolons are outside closing quotation marks.
- Commas and periods are inside closing quotation marks.

8.12 Semicolon

Semicolons are primarily used in long sentences to improve clarity and to control the pace of presentation. Use a semicolon where a brief pause is needed; the pausing effect of the semicolon is between the effects of a comma and a period.

A semicolon separates two independent clauses when they are of equal importance. In complex sentences, semicolons may precede transitional elements, conjunctive adverbs, or coordinating conjunctions.

Example 8-25. Semicolons separating elements of sentences

Transitional words or phrases:

The basics of heat transfer is important to the design of key components of powerplants; for example, boilers, economizers, and waste-heat recovery units.

Conjunctive adverbs:

The computational models are still running; therefore, we do not know whether they agree with the experiments.

COVID-19 has caused anxiety in Canada; in fact, the government reactions did not help much at the very beginning.

Coordinating conjunction:

The editorial office may invite qualified reviewers from academics, national labs, and large corporations; but not all of them will accept the invitations.

A semicolon may also separate elements in a series when one or more of the items contain commas. It avoids the confusion and achieve smooth flow of thoughts. Do not follow a semicolon with *and* or *or*, although you can do so for comma (Example 8-26).

Example 8-26. Semicolons separating items in a series

Semicolon: The co-authors of this article are Bin Zhao, Tsinghua University; Chun Chen, Chinese University of Hong Kong; Chao Tan, University of Waterloo.

Comma: The co-authors of this article are Bin Zhao, Chun Chen, and Chao Tan.

~~~

After revising the entire document three times, you should be satisfied with the language and the technical contents in your writing. Meanwhile, you may have formatted most part of the document based on your earlier experience while you were revising your document. Note that you formatted the document with distractions because your attention was focused on the language and technical contents. You now need to revise your document one more time, concentrating only on format.
~~~

9 Final Formatting

If outlining a document is like framing a house, then formatting the document is like furnishing the house. If you have paid one million dollars for the house, you surely expect it to look nice and to provide you with convenience too. Similarly, after you have put so much effort to drafting and editing your document, now it is time to bring its appearance and readability to the highest standard possible.

Most publishers provide the authors with guidelines on manuscript format. The guidelines include step-by-step instructions on formatting titles, headings, abbreviations, symbols, captions, visuals, equations, body text, and so on. Follow their instructions closely if you have identified the publishers; otherwise, you may follow the general guidelines described in this chapter.

9.1 Formatting the Title

Spell out the words whenever it is possible, and capitalize the following elements in the titles.

- First word
- Last word
- Initial letters of the major words for long documents
- Long prepositions with *five* or more letters (such as *between*, *since*, *until*, and *after*).

Except fully capitalized titles, do not capitalize words for short documents (articles, papers, *etc.*). Nor to capitalize coordinating conjunctions or short prepositions.

Example 9-1. Title format of a thesis

Incorrect: Recording clinical-grade ECG from the upper arm.

Correct: Recording Clinical-grade Electrocardiogram from the Upper Arm.

9.2 Formatting the Table of Contents

Most Tables of Contents (TOCs) can be automatically created by modern word processing software. The software ensures that the major *headings* and *subheadings* of your document are included in the TOCs with default formats.

There are many TOC styles available from the word processing software. Each style is designed for a specific group of readers. You can choose the right style to start with and refine it as needed. You may consider formal style for a formal document, like the one used in this book. It looks conservative, but professional.

9.3 Formatting Body Texts

9.3.1 Justification of margins

Justification of margins is a personal choice. Some like left-justified margins, and others prefer fully justified margins. Fully justified text appears formal, polished, and important. On the other hand, left-justified margins avoid uneven space between words. Furthermore, left justification is easy to process for software like Microsoft Office. Avoid using center alignment of paragraphs in the body, although headings may be center aligned, and visuals are normally center aligned. (*See* Example 9-2).

Example 9-2. Justification of margins

Left alignment (white spaces on the right)

Refrain from stretching or condensing words to fit them into one line. It distorts the letters, numbers, punctuation marks, and so on. Instead, slightly adjust the text by scaling up or down proportionally to fit the words into one line.

Full alignment (large space between words)

Refrain from stretching or condensing words to fit them into one line. It distorts the letters, numbers, punctuation marks, and so on. Instead, slightly adjust the text by scaling up or down proportionally to fit the words into one line.

Center alignment (awkward)

The spacing between lines is important to the legibility and comfort of reading. The spacing between lines should be just right, either too tight or too loose distorts the page layout.

Refrain from stretching or condensing words to fit them into one line. It distorts the letters, numbers, punctuation marks, and so on. Instead, slightly adjust the text by scaling up or down proportionally to fit the words into one line.

9.3.2 Line spacing

Line spacing is important to the legibility and comfort of reading. The space between lines should be just right, either too tight or too loose distorts the layout of the pages.

Apply the same line spacing to the entire document. The line spacing should range from 1-1.5 lines. You may use double-line spacing for drafting and editing, but not the final document. However, minor adjustments are acceptable to fit contents into one page.

Example 9-3. Line spacing

Normal spacing (1.2 lines):

Read through the document and check whether any heading or sub-heading is located at the bottom of a page, or any visual (including caption title) is separated onto two pages. You can fix these layout errors by carefully adjusting the line spacing.

Too tight (0.8 line):

Read through the document and check whether any heading or sub-heading is located at the bottom of a page, or any visual (including caption title) is separated onto two pages. You can fix these layout errors by carefully adjusting the line spacing.

Too loose (2 lines):

Read through the document and check whether any heading or sub-heading is located at the bottom of a page, or any visual (including caption title) is separated onto two pages. You can fix these layout errors by carefully adjusting the line spacing.

Read through the document and check whether any heading or sub-heading is located at the bottom of a page, or any visual (including caption title) is separated onto two pages. You can fix these layout errors by carefully adjusting the line spacing.

9.3.3 Typeface and type size

A typeface is the overall design of the letters used in your writing. It can be varied with **bold**, *italic*, *etc.* Your font changes with the typeface. There are many typefaces available, but you should use professional typefaces in technical writing that most readers are familiar and comfortable with. Commonly used typefaces in technical writing include Arial, Calibri, and Times Roman. Meanwhile, avoid strange typefaces that are illegible or childish (Example 9-4).

Example 9-4. Different typefaces

Avoid: Avoid typefaces such as **COMIC**, Bradley, and ***brush***.

Use: Use typefaces such as Arial, Calibri, Garamond, and Times New Roman.

Use consistent typefaces throughout a single document. However, you can use variants to achieve emphasis and contrast. For instance, distinctive typefaces are normally used for the titles, headings, and subheadings.

Most writers and publishers use 11 or 12 points as the default font size for the main text of their printed documents. The sizes of headings are normally larger than that of the body text, and headers and footers are typically smaller than the main text size. Otherwise, mismatched sizes look awkward.

9.3.4 Capitalization

There are numerous grammatical rules related to appropriate capitalization. To start with, you can refer to Table 9-1 for your writing. More can be found in handbooks for writing (*e.g.* Alred *et al.* 2018).

Example 9-5. Mismatch of font sizes and typefaces

Initial hydrogen ion concentrations for 5 ml, 10 ml, and 20 ml acid loading are 0.107, 0.199 and 0.355 molarity respectively. Based on the assumption of first-order reaction, the following equations can be obtained:

Biomass + Hydrogen ion → Product

For hydrogen ion loss: $r_H = -k_H\ [H^+] = d[H^+]/dt$

Reforming to get the expression equation for hydrogen ion concentration with first-order:

$$[H+] = [H+]_{initial}\, e^{-k_H * t}$$

Table 9-1. Capitalizations

Capitalization	Example
Address and places	Beijing, China; Ontario, Canada
Associations	Canadian Society of Mechanical Engineering
Beginning sentences	*[Do you need examples?]*
Corporations	Blackberry, BP, Microsoft Office
Cross-references	Chapter 5, Eq. 10, Section 2.3 Fig. 1, Table 6
Days and months	Monday, Sunday, January, December
Ethnic groups	Chinese, European
Historical events	World War II
Institutions	Tsinghua University, University of Waterloo
Internal units of corporations, *etc.*	Faculty of Engineering U.S. Department of Energy
Letters for shapes	*I*-beam, *T*-shape, *U*-turn
Names of people	John Lennon, Tom Smith
Nationalities	American, Arabic, Canadian, Chinese, German
Religions	Christianity, Jewish, Muslim

9.3.4.1 Don'ts in capitalization

Don't capitalize the first words when they are used as follows.

- The first word enclosed in dashes, brackets, or parentheses.

Example 9-6. Capitalization challenge (1)

Incorrect: We must repeat the tests (Data just collected are lost).

Correct: We must repeat the tests (data just collected are lost).

- The first word after a colon (:) or semicolon (;).

Example 9-7. Capitalization challenge (2)

Incorrect: We must repeat the tests; Data just collected are lost.
Correct: We must repeat the tests; data just collected are lost.

Incorrect: We must repeat the tests: Data just collected are lost.

Correct: We must repeat the tests: data just collected are lost.

- *Autumn*, *spring*, *summer*, and *winter* when they are used as seasons of the year.
- *Earth*, *sun*, *moon*, and the like except when they mean astronomical bodies.
- *East*, *north*, *south*, and *west* when are used for directions instead of places.

Example 9-8. Capitalization challenge (3)

Incorrect: Heading East, you will see a tall silver building.

Correct: Heading east, you will see a tall silver building.

9.3.4.2 Capitonyms

Capitonyms are dual meaning words, which change their meanings if the words are capitalized. Capitonyms may be challenging to some non-native English writers.

Table 9-2. An incomplete list of capitonyms

Capitalized		Lowercase	
Word	**Meaning**	**Word**	**Meaning**
China	the country	china	porcelain
Turkey	the country	turkey	the bird
March	the month	march	to walk (verb)
Titanic	the ship	titanic	gigantic (adjective)
Bill	name of a person	bill	amount to be paid
Lent	period in the Christina calendar	lent	past tense of *lend*
Reading	a town in England	reading	progressive form of *read*
Polish	relating to Poland	polish	smoothing by rubbing

9.3.5 Words and phrases with non-English roots

Italicize the words and the phrases with non-English roots; do so for their abbreviations too. Most of these words have Greek or Latin roots. Look up in dictionaries when you are in doubt, but the following are commonly used in Engineering writing.

Table 9-3. Words and phrases with non-English roots

Roots	Abbr.	Meaning	Usage in writing
exempli gratia	*e.g.*	for example	Usually enclosed in parenthesis; Examples can be found throughout this book.
et alii	*et al.*	and others	Normally used to end an incomplete list of names in citations and references. *See also* Section 9.12.2.
et cetera	*etc.*	and so forth	Used in incomplete lists in sentences. Examples can be found throughout this book.
id est	*i.e.*	that is	*See also* Example 4-52.
versus	*vs.*	against	Used to show relationship of one against the other (*e.g.*, Fig. 5e shows the change of density *vs.* pressure.)
vice versa	-	the other way around	Normally used to end a sentence following "*, and*" (, and *vice versa*). *See* Example 4-46.

9.3.6 Specifying items

Italicize the sentence elements (words, phrases, symbols, and so forth) when they are specified. You see italicized sentence elements throughout this book.

Example 9-9. Italicizing specified elements in sentence

The word *italicize* means to *use italics*; it is not the language or the country because the initial *i* is in lowercase.

9.4 Formatting Headings

You can be creative in heading styles, but they should be consistent throughout the document. Headings are not full sentences; therefore, make sure your headings are concise, but informative, in the final document. Avoid too many or too few words in the headings: too many words cause clutter and two few words fail in clarity.

Format the headings with consistence and agreement. Ensure that the font size, typefaces, and capitalizations at the same level agree with each other. However, the lower-level headings normally differ from upper level headings by size or style (such as bold face, capitalization, and italics).

Combinations of numbers and letters are used in most publications. Nonetheless, decimal numbering system appears in most technical reports, peer reviewed articles, and theses. (*See* Example 9-10)

Create a typographic contrast between headings and the text. The contrast improves clarity and emphasis. Effective contrast is normally achieved by using large font and bold face. All heading decrease font size in accordance to their levels. For this reason, make sure that the font size of the lowest-level headings equals that of the body text.

You can use the following check list to check your final formats.

- Same typeface for the same level of headings
- More numerical numbers in lower levels of headings
- More capital letters in higher levels of headings
- New typeface or smaller size at a lower level of headings

- Bolded face at the higher level of headings
- All headings flush left margin

Don’ts for headings:

- Don’t show part of a heading at the bottom of any page.
  Adjust line spacing to move the heading onto the next page.
- Do not indent headings.
  Decimal headings flush with the left margins in the final document, although they are usually indented in the TOC (Table of Contents).
- Don’t indent the first paragraph that follows the heading.
  Indent other paragraphs under the same heading or sub-heading.
- Don’t suppress headings.
  Instead, use double space for the headings.

Example 9-10. Formatted heading example

1. INTROUCTION

1.1 Literature Review

1.1.1 Motivation

1.1.2 Experimental works

1.1.3 Numerical works

1.2 Knowledge Gaps

1.3 Objectives

1.4 Thesis Structure

2. METHODOLOGY

2.1 Overall Experimental Setup

2.2 Test Apparatus

2.3 Instrumentation and Data Collection

2.4 Data Analyses and Expected Results

3. RESULTS AND DISCUSION

4. CONCLUSIONS

9.5 Formatting Headers and Footers

Use concise words, phrases, and clauses in headers and footers. You can avoid visual clutter by using concise elements using font sizes that are smaller than the main text.

Pay attention to the format of page numbers in footers. Page numbers in the front matter should be numbered consecutively with the Roman numerals using lowercase letters (*e.g.*, "i, ii, iii"). Differently, the page numbers of the main body and backmatter are numbered using Arabic numbers (*e.g.*, 1, 2, 3).

9.6 Formatting Dates

9.6.1 Format of dates

The formats of dates vary with countries. In the USA, dates are usually written in the format of *month day, year*. In Canada, however, dates are written in the pattern of *day month year* (*without commas in between*). On the contrary, dates follow the order *year month day* (nothing in between) in Chinese literature. Regardless of the format, a date normally follows the preposition *on*. When year is omitted, use the cardinal number (*January* instead of the ordinal number (*January 4th*).

Example 9-11. Format of dates (1)

USA:	The data were collected on January 4, 2020.
USA	The last day of this semester is August 30, 2020.
Canada:	The data were collected on 4 January 2020.
Canada:	The last day of this semester is 30 August 2020.

Example 9-12. Format of dates (2)

We submitted the manuscript to the editorial office on January 4.
The target date of re-submission is February 30.

9.6.2 Don't's in dates

Don't use only numerical form for dates in formal writing. Otherwise, it creates confusion or ambiguousness, especially to international readers. Depending on countries, *10/2/20* may be interpreted as *October 2, 2020*; *February 10, 2020*; *2 October 2020*; *10 February 2020*. Spelling out the month is important to clarity.

Example 9-13. Format of dates (3)

Incorrect: We submitted the revised manuscript on 10/2/20.

Clear: We submitted the revised manuscript on October 2, 2020.

Clear: We submitted the revised manuscript on 2 October 2020.

Clear: We submitted the revised manuscript on 10 February 2020.

Clear: We submitted the revised manuscript on February 10, 2020.

Don't use commas, however, when days are not shown in the dates. The preposition *in* may precede dates without days.

Example 9-14. Format of dates (4)

The data were collected in January 2020.

The target date of submission is in August 2020.

9.7 Formatting Numbers

9.7.1 Numerals or words for numbers

Many non-native English writers use numerals (1, 2, 100) for numbers in their writing, which is informal in English writing. In the text of formal writing, you need to write words for *integral* numbers ranging from zero through ten (*one, two, … ten*) and use numerals for the larger numbers. Use commas to separate the elements of Arabic numbers greater than 999 using the format 1,234,567,890.

However, always use numerals for numbers in equations, with decimals (3.14, 1.618), and with units.

Example 9-15. Format of numbers in sentences (1)

Informal: I have published 3 books and 80 articles, and my goal is to publish 10 books and 150 articles in total before I retire.

Formal: I have published three books and 80 articles, and my goal is to publish ten books and 150 articles in total before I retire.

For sentences beginning with numbers, use words for the numbers. You can also reconstruct the sentences to avoid starting with numbers.

Example 9-16. Format of numbers (2)

Incorrect: 400 international attendees participated in the annual event.

Correct: Four hundred international attendees participated in the annual event.

Revise: This year, 400 international attendees participated in the annual event.

Spell out one of them for clarity when one number precedes another in the same phrase.

Example 9-17. Format of numbers (2)

Confusing: Our samples include 12 6-meter long steel bars.

Revised: Our samples include twelve 6-meter long steel bars.

Revised: Our samples include 12 steel bars, which are 6 m long.

Revised: Our samples include 12 six-meter long steel bars.

9.7.2 Formatting time

Time may be occasionally written into technical documents to specify the exact time of action. It may be an important factor for research like atmospheric air dispersion or solar energy.

Colons are used to separate hours from minutes in time followed by abbreviations *a.m.* or *p.m.* (for example, 8:30 a.m., 12:30 p.m.) You need to spell out the words of time when *a.m.* or *p.m.* is not included. (*e.g.*, eight thirty in the morning; 12:30 in the afternoon; Seven o'clock in the evening).

9.8 Formatting Lists

Lists are used more frequently in longer documents. There are lists in almost all technical books, dissertations, formal reports, and the like. However, you should be prudent with lists if page limits are imposed on your writing. For instance, some contents in Example 4-52. Well-written methodology could be listed if there were no page limit set by *Analytical Chemistry*.

When you format a list, check carefully to make sure that the entire list maintains parallel structure and format throughout. All listed items normally begin with capital letters; listed sentences should also end with period marks.

- Number and letter lists

  Use numbers to indicate sequence or rank. Check that the items are listed in the order you intended. A combination of numbers and letters allows you to list with subdivisions.

- Bullet points

  Bullets are used for items without clear rank or sequence. However, you may still want to list the items in a logical order that supports your purpose of writing.

Don'ts for lists

- Don't end items with commas or semicolons.
- Don't use the word *and* or *or* at the to join consecutive items.

9.9 Formatting Units and Symbols

Use the International System of Units (abbreviated SI) in your writing unless you are clearly instructed otherwise. SI units are widely used in scientific and engineering publications, especially those aimed at international readers. Meanwhile, more and more international publishers are encouraging the usage of international measurement standards.

The following grammatical rules of unit are generally applicable internationally.

- Leave one space between the measurement value and its unit.
- Use a small dot (·) or period (.) between two units to indicate multiplication operation.
- Use a backslash (/) between two units to indicate division operation.

- Do not follow an abbreviated unit with a period mark (kg, kPa).
- Do not add letter "*s*" to an abbreviated unit for plural.
- Capitalize the abbreviated units named after people (*See* Table 9-4); otherwise, the abbreviated units are written in lowercase.

Table 9-4. Symbols and units named after pioneer researchers

Symbol	Unit	Named after	Used for
A	ampere	André-Marie Ampère	Electric current
C	coulomb	Charles-Augustin de Coulomb	Electric charge
°C	degree Celsius	Anders Celsius	Temperature
Hz	hertz	Heinrich Rudolf Hertz	Frequency
J	joule	James Prescott Joule	Energy
K	kelvin	William Thomson Kelvin	Temperature
N	newton	Isaac Newton	Force
Ω	ohm	Georg Simon Ohm	Electric resistance
Pa	pascal	Blaise Pascal	Pressure
V	volt	Alessandro G. A. Anastasio Volta	Electric potential (voltage)
W	watt	James Watt	Power

9.10 Formatting Visuals

Basics of creation and format of visuals has been introduced in Section 5.4. However, you might have changed their formats and locations while revising the documents. You need to relocate certain visuals and refine their formats before proofreading.

9.10.1 Placement of visuals

First, check and ensure that all visuals are placed with center alignment and that their captions are placed with consistence: figure captions are under the figures and table captions are above the tables. Then, make sure that all the caption titles follow the same rules of grammar, spelling, and format.

Check and ensure that visuals are integrated into the body text, and they appear right after their first explanations. All visuals should be cross-referenced in the text, and *vice versa*. You may move paragraphs around while revising the entire document to further improve the coherence and unity. Now, check the relative locations of the explaining text and the visuals before finalizing your document for proofreading.

Occasionally, visuals are gathered in one place in manuscripts under review. For example, many international journals require authors to submit their manuscripts with figures and tables appended to the text-only documents, which contain only visual captions. However, that is only for the ease of peer review and editorial tasks. In the final document, visuals normally appear right after their first cross-references.

9.10.2 Formatting visuals

How you format visuals reveals your skills in visual design: it is personal. Appealing visuals improve the reading experience, which can positively impact your career advancement. So, format your visuals as if you were an artist. Admittedly, engineers do not pretend to be artistic designers, but the following basics in visual design may help you improve the effectiveness of the visuals you created. With enough attention and continual practice, you should become more and more skilled in visuals for engineering publication.

As explained in 5.4, visuals are composed of basic elements such as colors, forms, lines, shapes, texts, and white space. Their forms may be one-, two- or three- dimensional. By considering all these factors, your formatting improves the clarity of the main document.

Maintain a balance between consistence and variety of visuals; The variety of visuals improves clarity. Variations in color and font size in the same visual create contrast and hierarchy effects, emphasizing the important elements.

Do not clutter your visuals. Leaving enough spaces around lines, shapes, and texts increases readability. White space is important to the comfort of eyes too. You can increase comprehension of visuals by proper use of white spaces between visual elements and within individual elements. You are encouraged to read some books about the white space in visual design (*e.g.*, Hagen and Golombisky, 2013).

You can also find many nicely crafted visuals in the public domain. Examine them against the general guidelines introduced in this book and apply them to your writing. Nonetheless, you may also learn from the following negative examples. (*See* Section 5.4.2 for more visuals.)

Figure 9-1 is carelessly formatted figure. Without labels, the readers cannot understand the figure without reading the related text – then the figure is meaningless. In addition, this figure is not clear because of the blurry and unnecessary lines and shapes.

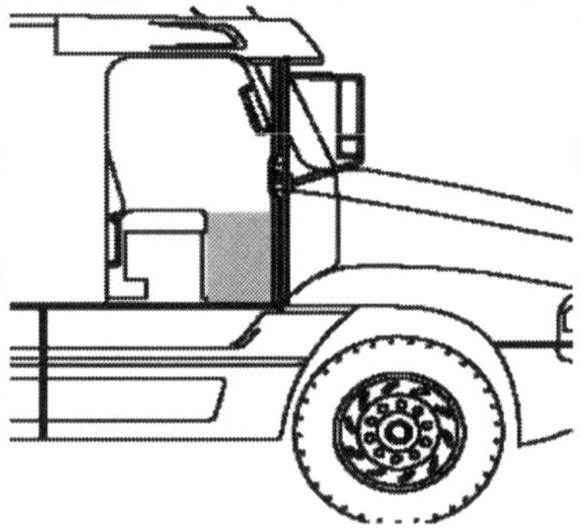

Figure 9-1. Blurry figure missing text

Figure 9-2 shows a figure of an article that is published in a prestigious international journal (Zhou *et al.,* 2011). This figure has both pros and cons. On one hand, the figure is produced with clarity: lines, shapes, and texts are properly laid out. On the other hand, it is not clear what the symbols stand for, although there is enough room for explanatory words. Another problem is the font size mismatch between the body text and the caption title of the figure. This mismatch can be easily fixed, should the authors follow the basic guidelines about visual formatting.

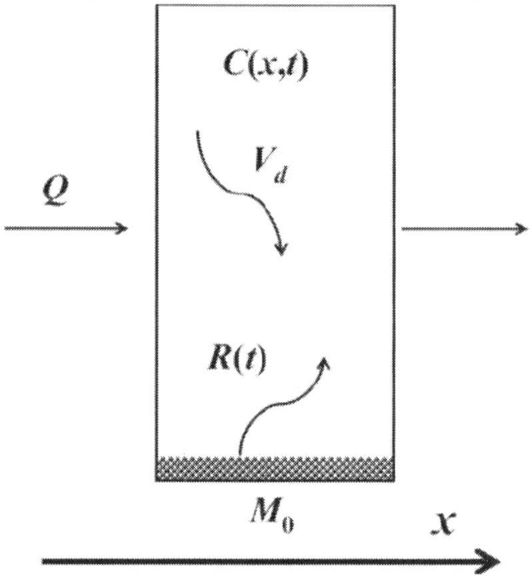

FIG. 1. Illustration of the process involving airflow, particle resuspension, and deposition in a duct unit.

Figure 9-2. Poorly formatted figure with font size mismatch

9.11 Mathematical Equations

Formatting equations is a simple but tedious work. The following formatting guidelines are applicable for all equations. Number the equations consecutively throughout the final document. Most writers enclose the equation numbers with parentheses or brackets, which are on the right side of the equations and flush to the right margin. Leave at least three spaces between the equations and their numbers to avoid misinterpretation. Indent all the equations by at least half an inch from the left margin: do not type the equations from the left margin.

In the body text, the equations are cross-referenced by equation numbers preceded with the word *Equation* or the abbreviation *Eq.*, which is capitalized (*See also* Table 5-1). Either one is acceptable, but it should be used consistently throughout the document. However, do not list equations in the text unless the equation itself is part of conclusions. Table of Equations is not needed in your document either.

As illustrated in Example 9-18, a long equation that cannot fit into one single line should be broken into two or more lines. The new lines should start with an equal sign or an operation sign (+ − × ÷) that is *not* enclosed with parentheses or brackets. Align the left side of the first line and right side of the last line with other single-line equations on the same page. The lines in between can be centered or aligned by the right of equations. The long equation is numbered as one equation.

Example 9-18. Alignment and arrangement of a long equation

The mass balance in the cubic volume gives

$$\Delta C(\Delta x \Delta y \Delta z) = \left(\Delta J_x + \Delta J_y + \Delta J_x\right)\Delta t \tag{9-1}$$

Substituting Equation (x-y) into Equation Eq. 9-1 you can get

$$\begin{aligned}\frac{\Delta C}{\Delta t} = \frac{(C_x - C_{x+\Delta x})u + \left[\left(-D_x \frac{\partial C}{\partial x}\right)_x - \left(-D_z \frac{\partial C}{\partial z}\right)_{x+\Delta x}\right]}{\Delta x} \\ + \frac{\left[\left(-D_y \frac{\partial C}{\partial y}\right)_y - \left(-D_y \frac{\partial C}{\partial y}\right)_{y+\Delta y}\right]}{\Delta y} \\ + \frac{\left[\left(-D_z \frac{\partial C}{\partial z}\right)_z - \left(-D_z \frac{\partial C}{\partial z}\right)_{z+\Delta z}\right]}{\Delta z}\end{aligned} \tag{9-2}$$

Considering the limit of an infinitesimally small cube and time interval, Eq. 9-2 becomes

$$\frac{\partial C}{\partial t} = -u\frac{\partial C}{\partial x} + D_x \frac{\partial^2 C}{\partial x^2} + D_y \frac{\partial^2 C}{\partial y^2} + D_z \frac{\partial^2 C}{\partial z^2} \tag{9-3}$$

9.12 Formatting References and In-text Citations

There are three principal formatting systems for in-text citations and references: American Psychological Association (APA) system (APA, 2020), Institute of Electrical and Electronics Engineers (IEEE) system (IEEE, 2020), and Modern Language Association (MLA) system (MLA, 2020). APA and IEEE styles or their combination are normally used in engineering publications.

The standard formatting styles are meant to be guidelines and they should not discourage your creativity. Be yourself within certain constraints; so is true for writing. Regardless of the styles you choose, they should be consistent throughout your document.

9.12.1 References

The page layout of the *References* is the same as the document body. The line spacing is also the same as that of the body text. (*See also* 9.3.) The heading *References* appears on a new page in the final document. There may be exceptions for short documents such as articles in international journals and extended abstracts in the proceedings of international conferences.

References are listed without bullet or numerical numbers. Treat each entry of reference list as one single paragraph. Align the first lines of the entries to the left margin but indent one-half inch from the left margin for the other lines to follow. *See* Example 9-19.

Example 9-19. Layout of reference list

Dolan R, Yin S, Tan Z. 2009. Design and evaluation of a subcritical hydrothermal gasification system. *Proceedings of 50th International Conference on Bioenergy*, December 1-3, 2009, Calgary, Canada.

Dolan R, Yin S, Tan Z. 2010. Effects of headspace fraction and aqueous alkalinity on subcritical hydrothermal gasification of cellulose. *International Journal of Hydrogen Energy* 35:6600-6610.

As seen in *References* of this book, the entries are listed in an alphabetical order by the first (or sole) author's last name. Multiple entries written by the same author(s) are listed in the order of publication time, from early to recent. List each author's name by last name, followed by initial(s). More details are listed as follows.

- List all authors following the same pattern.

  Separate authors with commas. Stop at the sixth author (inclusive) for papers with more than six authors. Do not include the seventh author. Instead, use *et al.* to represent the rest of the authors.

- Capitalize only the first words for the titles of articles, conference papers, and the like.

- Use lowercase without Italicism for titles of short documents, such as (book chapters and journal articles).
- Capitalize and italicize the titles of long documents (books, theses, and periodicals) (*See also* 4.6)
- Separate the title and the subtitle in the same entry using a colon.
- Do not enclose any title or subtitle with quotation marks, parentheses, or the like.
- Separate the list of authors, year, title, page range, and other elements using commas or periods.

Last, but not the least, check and ensure that each entry of the reference in the list is cited in your text; likewise, each in-text citation corresponds to one and only one reference. However, bibliography does not require this matching because there should be more bibliographic entries than their in-text citations.

9.12.2 In-text citations

By in-text citations, which appear in the main text, you give credit to others for their contributions to your writing. There are at least three forms of in-text citations: number in brackets, number as superscript, and *(author, year)* form. Most publishers use the *author-year* form for the upfront credit to the author(s), but the citations in parentheses are somehow interruptive to the smoothness of writing and reading.

Pay attention to the difference between parenthetical citation and narrative citation. The name of the author is part of the sentence when narrative citation is used, whereas parenthetical citations can be removed from the sentence without affecting the ideas of the sentence. (*See also* Example 9-20 for illustration of difference.)

The exact form of in-text citation depends on the number of authors. Single authored work takes (author, year) format by default (*e.g.* Tan, 2005). For a work with two authors, cite both names joined by the word *and* (for example, Tham and Roy, 2020). For a work with three or more author, list only the first name followed by *et al.* (for example, Tan *et al.*, 2014). When you enclose multiple citations using the same parenthesis, list the them alphabetically and separate them with semicolons (Adam, 1971; Tham and Roy, 2020; Tan *et al.*, 2014).

Example 9-20. Parenthetical citations and narrative citations

Parenthetical citations:

Other researchers have reported the effects of headspace fraction and aqueous alkalinity on subcritical hydrothermal gasification of cellulose (Dolan *et al.*, 2010). *[reduces interruption]*

Other researchers (Dolan *et al.*, 2010) have reported the effects of headspace fraction and aqueous alkalinity on subcritical hydrothermal gasification of cellulose. *[Position in between interrupts the flow of reading.]*

Figure 5-5 shows a figure in the article recently published in the Journal of Power Sources (Li *et al.*, 2019).

Narrative citation:

Dolan *et al.* (2010) reported the effects of headspace fraction and aqueous alkalinity on subcritical hydrothermal gasification of cellulose.

Figure 5-5 shows a figure in the article written by Li *et al.* (2019).

9.12.3 Recommended styles for references and citations

No single standard style pleases all readers. I recommend the following styles for long documents and short documents.

- **Articles in journals, magazines, *etc.***
  Author(s). Year. Title of article, *Title of Journal* Volume number (Issue number): beginning page number – last page number.
- **Peer-reviewed conference papers**
  Author(s). Year. Title of paper, *Proceedings of the Conference Name*, Conference time, Location.
- **Chapters in books and reports**
  Author(s). Year. Chapter title, in *Title of Long Document*, edition, Publisher Name, Address.
- **Works in collections:**
  Author(s). Year. Chapter title, in *Title of Long Document* (editor: name), edition, Publisher Name, Address.
- **Books and large reports**
  Author(s). Year. *Title of Document*, edition, Publisher, Address.
- **Degree theses**
  Author(s). Year. *Title of Document*, Doctoral thesis or Master's thesis, University Name, Country.

Table 9-5 summarizes my recommendations for reference entries and the corresponding parenthetic in-text citations. They are primarily based on APA system, but with some modifications to improve conciseness and clarity.

9.12.4 Formatting cross-references

A cross-reference directs readers to another place for related text in the same document. Section numbers and page numbers are optional cross-references, but they help the readers find the related text quickly. You do not have to include page numbers, and you can use *See* or *See also* to introduce cross-references.

Make sure all cross-references are formatted consistently throughout the document. It is optional to enclose cross-references in parentheses. Italicize the word(s) *See* or *See also*. When page number is included, insert a comma in front of the page number.

Table 9-5. Styles of reference entries and in-text citations

Type	Author	Example	Citation
Journal article	Single author	Tan Z. 2008. An analytical model for the fractional efficiency of a uniflow cyclone with a tangential inlet, *Powder Tech.* 183 (2): 147-151.	Tan, 2008
	Two authors	Givehchi R, Tan Z. 2015. Effects of capillary force on airborne nanoparticle filtration, *Journal of Aerosol Science* 83: 12–24.	Givehchi and Tan, 2015
	Three to six authors	Li J, Cheng K, Croiset E, Anderson WA, Li Q, Tan Z. 2017. Effects of SO_2 on CO_2 capture using chilled ammonia solvent, *International Journal of Greenhouse Gas Control* 63: 442-448.	Li *et al.*, 2017
	Seven or more authors	Johnson MD, Fish VL, Doeleman SS, Marrone DP, Plambeck RL, Wardle JFC, *et al.* 2015. Resolved magnetic-field structure and variability near the event horizon of sagittarius A, *Science* 350: 1242.	Johnson *et al.* 2015

Table 9-5. (continued)

Type	Author	Example	Citation
Paper	Same as above	Sun F, Sade B, Ghosh H, Tan Z, Sivoththaman S. 2019. Synthesizing reduced graphene oxide, *Proceedings of the 46th IEEE Photovoltaic Specialist Conference*, June 16-21, 2019, Chicago, IL, USA.	Sun *et al.* 2019
Work in collect-ion	Same as above	Tan Z. 2013. Nanoaerosol, in *Encyclopedia of Microfluidics and Nanofluidics* (ed. Li D), 2nd ed., Springer Verlag, Singapore.	Tan, 2013
Collect -ion	Same as above	Tan Z. 2019. *Micro/Nano Materials for Clean Energy and Environment*, special issue of *Materials*, MDPI, Basel, Switzerland.	Tan, 2013
Book	Same as above	Tan Z. 2014. Air Pollution and Greenhouse Gases, Springer Verlag, Singapore.	Tan, 2014
	Same as above	Ashrafizadeh SA, Tan Z. 2018. *Mass and Energy Balances*, Springer Verlag, New York, USA.	Ashrafizadeh and Tan,2018
	Corporate author	Mathworks (2020). *Matlab Numerical Computing Tutorials.* Natick, MA, USA.	Mathworks, 2020
	Govern-ment author	DOE (Department of Energy). 1995. *Life-Cycle Costs for the Department of Energy Waste Management Programmatic Environmental Impact Statement* (Report INEL-95/0127 [Draft]). Department of Energy, Washington DC, USA.	DOE,1995
Thesis	Normally single author	Saprykina A. 2009. *Airborne Nano-particle Sizing by Aerodynamic Particle Focusing and Corona Charging.* MSc thesis, University of Calgary, Canada.	Saprykina, 2009
		Givehchi R. 2015. *Filtration of NaCl and WOx Nanoparticles Using Metal Wire Screen and Nanofibrous.* PhD dissertation, University of Waterloo, Canada.	Givehchi, 2015

9.13 Formatting Appendices

Format the appendices following the same styles as in the body text. If you have only one appendix, label it as *Appendix*. If you have more than one appendix, give each of them a heading. (*Appendix A*, *Appendix B, etc.)* Do not use numerals in the appendix headings; instead, use letters *A*, *B*, *C*, and so on. You may generate a List of Appendices like the List of Figures for a long document. *See* Appendix in this book for example.

9.14 Formatting Index

Index is unnecessary to short documents, and it is optional to long documents. Many books or large reports have indexes, but I rarely found them useful. In addition, it would be time-consuming to manually create the indexes of your long documents. Nowadays, most word processing software allows you to create indexes automatically. To start with, you may refer to *The Chicago Manual of Style* by the University of Chicago Press Editorial Staff (2017) for guidelines on index.

This book is relatively short, and it does not require an index. As an example, however, a simple index is created as an example only (*See* the last page of the book).

~~~

You must ask all the co-authors, if any, to contribute to revising the document. They may participate at different stages or different depths, but it should be prior to proofreading. By now, your manuscript is ready for proofreading. It is beneficial to ask people who are familiar with the field to proofread your document. They may give you unexpected but valuable recommendations for further improvement.
~~~

10 Proofreading and Others

10.1 Proofreading

Proofreading is the final step in academic writing for publication. Do not rush into proofreading; you should proofread only after you have finished all other writing tasks. Your proofreading is meant to focus on surface errors in spelling, grammar, punctuation, and format. Meanwhile, you can identify and refine the imperfections in typefaces, spaces, alignments, and visuals.

You can begin proofreading with spelling and grammar with the word processing software or other online tools. These tools, however, have limited database and rules. They can identify the spelling and grammatical errors, but not the way you develop and present your ideas. The machine cannot understand your tone, voice, logic, *etc.*: these are factors that characterize your writing styles.

Make sure your final document looks impressive, although impression is secondary to the clarity, coherence, and clarity of the writing. Like it or not, the way your document appears affects readers' interests. Your readers judge your professionalism based on your writing. A good impression matters. You may never know who is reading your document, but your effort in proofreading pays off in the end.

Be patient and proofread slowly.

10.2 Plagiarism Free and Copyright Clearance

Proofread and make sure that your document is free of plagiarism. Software such as *Turnitin* and *iThenticate* is a useful tool for plagiarism checking. While *Turnitin* is primarily for educational purpose, *iThenticate* is aimed at researchers to ensure the originality of scholarly publication. Ask your employer for the site license. Both *Turnitin* and *iThenticate* can generate similarity reports, which identify the possible plagiarism in your writing.

You must obtain written permissions from the copyright holders to use copyrighted texts and visuals. You can request the permissions by contacting the individual copyright holders. You may submit your request forms online if the copyright holders are larger corporations like *Springer* and *Elsevier*. One way or another, you need to deal with the permissions on a case by case basis.

10.3 Securing Internal Clearance

You need to secure your employer's permission to publish any work that may impact their business. Be careful with research funded by sponsors who may have signed a legal document, which you may not know, with your employer. On campus, ask your academic supervisor, who are likely your co-author, for final review before submission. Give a copy of the manuscript to your supervisor for review before sharing it with anyone external or submitting it anywhere for the consideration of publication.

You must secure approvals for publication from all co-authors too. It would be unethical to list a co-author, who is usually influential, without his or her approval. This may sound funny, but it did happen multiple times to me and to some of my colleagues. Regardless of the intention or excuses, such behavior is not acceptable. Remember, you are expected to be professional in academic writing for engineering publication.

Now your document is ready!

References

Alred GJ, Brusaw CT, Oliu WE. 2018. *Handbook of Technical Writing*, 12th edition, Bedford/st Martins, Boston, MA, USA.

APA (American Phycological Association). 2020. *Publication Manual of the American Psychological Association*, Seventh Edition. Washington, DC, USA. (https://www.apa.org/)

Bai Y, Yang K, Sun D, Zhang Y, Gao X. 2013. Numerical aerodynamic analysis of bluff bodies at a high Reynolds number with three-dimensional CFD modeling, *Science China Physics* 56: 277–289.

Cheng L. 2010. *Lignin Degradation and Dilute Acid Pretreatment for Cellulosic Alcohols Production.* MSc thesis, University of Cincinnati, Ohio, USA.

Dolan R, Yin S, Tan Z. 2010. Effect of headspace fraction and aqueous alkalinity on subcritical hydrothermal gasification of cellulose. *International Journal of Hydrogen Energy* 35: 6600-6610.

Du W, Bao X, Xu J, Wei W. 2006. Computational fluid dynamics (CFD) modeling of spouted bed: Assessment of drag coefficient correlations, *Chemical Engineering Science* 61 (5): 1347-1740.

EPA (Environmental Protection Agency). 2020. *2019 Automotive Trends Report: Greenhouse Gas Emissions, Fuel Economy, and Technology Since 1975.* Report number: EPA-420-R-20-006 (March 2020).

Fuller SJ, Zhao Y, Cliff SS, Wexler AS, Kalberer M, 2012. Direct surface analysis of time-resolved aerosol impactor samples with ultrahigh-resolution mass spectrometry, *Analytical Chemistry* 84 (22): 9858-9864.

Ge H, Hu D, Li X, Tian Y, Chen Z, Zhu Y. 2015. Removal of low-concentration benzene in indoor air with plasma-MnO_2 catalysis system, *Journal of Electrostatics* 76: 216-221. (doi.org/10.1016/j.elstat.2015.06.003)

Givehchi R. 2015. *Filtration of NaCl and WOx Nanoparticles using Metal Wire Screen and Nanofibrous Filters.* PhD dissertation, University of Waterloo, Canada.

Gioia G, Chakraborty P, Gary SF, Zamalloa CZ, Keane RD. 2011. Residence time of buoyant objects in drowning machines, *Proceedings of the National Academy of Sciences of the United States of America* 108 (16):6361-6363.

Hacker D, Sommers N. 2018. A Writer's Reference, ninth edition. Bedford/st Martins, New York, NY, USA.

Hagen R, Golombisky K. 2013. *White Space is Not Your Enemy: A Beginner's Guide to Communicating Visually Through Graphic, Web & Multimedia Design.* Taylor & Francis Group, New York, USA.

IEEE (Institute of Electrical and Electronics Engineers). 2020. *IEEE Editorial Style Manual for Authors* (V 04.10.2020). IEEE Publishing Operations, Piscataway, NJ, USA.

Li Y, Li Q, Tan Z. 2019. A review of electrospun nanofiber-based separators for rechargeable lithium-ion batteries, *Journal of Power Sources* 443: 227262.

Liu C, Wang C-C, Kei C-C, Hsueh Y-C, Perng T-P. 2009. Atomic layer deposition of platinum nanoparticles on carbon nanotubes for application in proton-exchange membrane fuel cells, *Small* onlinelibrary.wiley.com/doi/abs/10.1002/smll.200900278

Min J. 2015. *Quantifying the Effects of Winter Weather and Road Maintenance on Emissions and Fuel Consumptions.* Master's Thesis, Univ of Waterloo, Canada.

MLA (Modern Language Association). 2020. *MLA Handbook*, eighth edition. MLA, New York, NY, USA (mlahandbook.org)

Siddiqi MA, Petersen J, Lucas K. 2001. A study of the effect of nitrogen dioxide on the absorption of sulfur dioxide in wet flue gas cleaning processes, *Industrial & Engineering Chemistry Research* 40 (9): 2116-2127.

Tan Z. 2014. *Air Pollution and Greenhouse Gases: from Basic Concepts to Engineering Applications for Air Emission Control.* Springer Verlag, Singapore.

The University of Chicago Press Editorial Staff. 2017. *The Chicago Manual of Style*, 17th edition. University of Chicago Press, Chicago, IL, USA. (DOI: 10.7208/cmos17)

Zhou B, Zhao B, Tan Z. 2011. How particle resuspension from inner surfaces of ventilation ducts affects indoor air quality - a modeling analysis, *Aerosol Science and Technology* 45: 996-1009.

Appendix

Transitional Words and Phrases (Hacker and Sommers, 2018)

To compare:

also, in comparison, in the same manner, likewise, similarly.

To contrast:

although, and yet, at the same time, but, despite, even though, however, in contrast, in spite of, nevertheless, on the contrary, on the other hand, still, though, yet.

To give examples:

for example, for instance, in fact, specifically, that is, to illustrate, as an illustration

To indicate logical relationship:

accordingly, as a result, because, consequently, for this reason, hence, if, otherwise, since, so, then, therefore, thus

To indicate sequence:

first, second, third, initially, then, next, finally

To show addition:

additionally, again, and, also, besides, equally important, further, furthermore, in addition, in the first place, moreover, next, too

To show place or direction:

above, below, beyond, close, elsewhere, farther on, here, nearby, opposite, to the left (right, south, north. *etc.*)

To show time order:

after, afterward, after that, as, as long as, as soon as, at last, before, before that time, during, earlier, finally, formerly, immediately, later, meanwhile, next, now, since, since then, shortly, subsequently, then, thereafter, until, when, while

To summarize or conclude:

all in all, in conclusion, in other words, in short, in summary, on the whole, that is, therefore, to sum up

Index

Made in the USA
Coppell, TX
05 July 2020

29856567R00118